The Greatest Explosions in History

Fire, Flash, and Fury

Also by Ragnar Benson:

Action Careers
Breath of the Dragon: Homebuilt Flamethrowers
Bull's Eye: Crossbows by Ragnar Benson
Gunrunning For Fun and Profit
Hard-Core Poaching
Homemade C-4: A Recipe for Survival
Live Off the Land in the City and Country
Mantrapping
The Most Dangerous Game:
  Advanced Mantrapping Techniques
Ragnar's Guide to Home and Recreational
  Use of High Explosives
Ragnar's Ten Best Traps . . . And a
  Few Others That Are Damn Good, Too
Survivalist's Medicine Chest
Survival Poaching
The Survival Retreat
Switchblade: The Ace of Blades

The Greatest Explosions in History

Fire, Flash, and Fury

by Ragnar Benson

A CITADEL PRESS BOOK
Published by Carol Publishing Group

First Carol Publishing Group Edition 1991

A Citadel Press Book
Published by Carol Publishing Group
Citadel Press is a registered trademark of Carol Communications, Inc.

Editorial Offices	Sales & Distribution Offices
600 Madison Avenue	120 Enterprise Avenue
New York, NY 10022	Secaucus, NJ 07094

In Canada: Musson Book Company
A division of General Publishing Co. Limited
Don Mills, Ontario M3B 2T6

First published as *Fire, Flash, and Fury* by
Paladin Press, Boulder, Colorado.

Manufactured in the United States of America
ISBN 0-8065-1278-4

10 9 8 7 6 5 4 3 2 1

Contents

Preface

There are those among us to whom the smell of powder and the feel of the blast have become an addiction. The sense of elation in a high-explosive situation is difficult for most people to articulate. It could be a sense of power, of awe, or perhaps that massive explosions are a sign of dominance. On close questioning, most addicts will admit that this attraction to an earth-rending explosion is completely irrational. To be sure, nitroglycerine fumes will make one's heart race under any circumstances. This is not emotional but medical. Nitroglycerine pills are commonly used to treat heart patients.

I first realized that I was chemically unlike many of my peers while riding around in an open basket on an M60-A3 tank. Every time the crew discharged the main gun, the blast about dusted me off the machine. It was as though all our sixty-three tons had hit a wave of incoming jello; ever so positively and forcefully, it mushed the giant steel machine back. Were it not for the locked-down, braced position I assumed, legs wedged against the steel straps, I would have been whooshed from the turret. Earplugs, earmuffs, a flak helmet, and goggles kept me from losing parts of my vital senses.

I loved every moment of it. The throaty-sounding, authoritative thump up through the ground that I could feel in the back of my mouth; the smoke, flash, and fire from a good, healthy blast . . . all generated a high that is virtually impossible to describe. I felt a sharp rush of wind as the super-high-velocity explosion from the small LAW warhead hit me in the face. We were 300 meters from the impact area, but the concussion easily carried that far. High-velocity explosives will slap a person. Our instructor at the LAW range told us that, psychologically, some people are unable to endure standing in the departure zone of a high-explosive rocket. "Not I!" I had said as I stepped up to fire one of the plastic monsters.

I still remember my first firecracker vividly. Uncle Ralph, our kid uncle, brought them by for our consideration. Uncle Ralph taught us quite a lot (and not just about fireworks). The smoke, flash, and shredded paper all made a lasting impression. We diligently kept at it, graduating to larger and larger firecrackers. We thought M80s were it till the time in Mexico, when I was able to buy a whole suitcase full of cojetes made of old newspaper and a lovely quarter pound of powder.

At thirteen, I went to work for the old powder monkey in our village. I saw a bit of the work serious powder could do but seldom heard a report or saw a flash. "Waste" was not part of the vocabulary of my explosives mentor. We did not load our shots past the amount needed to "do the work," as those in the trade put it.

Preface

Powder was cheap in those days. A ten-cent stick of 60 percent made an excellent target, Fourth of July salute, or theatrical adjunct. Soon I pitied those in high school who had to make do with mere firecrackers. My father thought shooting dynamite was a terrible waste till he tried it a time or two. (The last time I saw him, he was piling up six or eight sticks for a more awe-inspiring target.)

Most regular people don't understand—much less appreciate—us addicts. Some folks can't even imagine what it is we are high on. The ultimate high occurred almost twenty years ago, on December 23, 1969, in a little village called Moscow, Idaho, where I was visiting my daughter. Some dear soul managed to touch off twenty thousand pounds of ammonium nitrate all in one shot. I was walking down Main Street when the concussion cracked all the west-facing windows. It was truly wonderful. It then became obvious to me that large quantities of explosives are allowed to run free upon the land occasionally. Great holes have been torn in the earth. Clouds of debris have risen to threaten private aircraft. Impressive to those who see it happen, to say the least.

This book is dedicated to those hardy souls who share my enthusiasm for huge explosions.

The Greatest Explosions in History

Fire, Flash, and Fury

Volcano Krakatoa erupts; resulting tidal waves sweep 37,000 to death

KRAKATOA, August 1883—Depending on the depth of one's theology, Krakatoa was either the greatest accidental explosion in the recorded history of man or it was God's most powerful act of nature in recent times. No man-made explosion ever unleashed on this earth even came close to Krakatoa. The best nuclear blast was not equal to even a puny, anaemic, fraction of what happened to the tiny group of volcanic islands in Southeast Asia on August 27, 1883.

Dr. William Brown, a fellow at the Hudson's Institute, recently ran new estimates on the explosion. On the absolutely low end, he calculated that the blast equalled ten thousand megatons of pure TNT. Should one accept his high figure, the blast unleashed forces equal to thirty thousand megatons of TNT.

Hiroshima's bomb, by contrast, was a trifling 20 kilotons. Twenty kilotons of TNT can fit in a twelve-foot cylinder the length of a football field. Kilo, of course, refers to thousands. Mega, as in Krakatoa, refers to millions of tons of high explosive. A metric ton is equivalent to 2,204 U.S. pounds. Ten thousand megatons is equal to 10 billion tons. If one assumes the lower of Dr. Brown's estimates, we are looking at an explosion of more than 22 trillion pounds of TNT. A railroad train comprised of 166,666,667 boxcars could conceivably haul 10 billion tons of TNT. Such a train would stretch 1,893,939 miles, or around the earth's equator at least 79 times! These figures are so astronomical that the only thing that can put them in perspective is the U.S. national debt.

The intensity of the blast at Krakatoa was such that five cubic miles of volcanic rock was pounded instantly into flour-fine dust. Particles were boosted aloft to 120,000 feet and traveled around the globe. Scientists estimate that some of Krakatoa settled on every square

KRAKATOA—THERE SHE BLOWS—Photo on previous page shows the eruption of volcano Krakatoa the night of August 26, 1883. A small volcanic island between Java and Sumatra, Krakatoa was almost destroyed. Tidal waves following the eruption swept thirty-six thousand people to their deaths in the Dutch East Indies. Weather around the world was affected for three years after the explosion due to dust in the upper atmosphere. (Photo courtesy of The Bettmann Archive.)

foot of the earth's surface. Particulate matter remained in the stratosphere for up to three years, cooling the earth below. It took until 1886 for the dust to circle the globe and to affect the weather in the United Kingdom. That year, England experienced its worst winter and shortest growing season in history. By the last of December, uncharacteristically severe blizzards with deep snow swept across the land. Frosts lasted through March. Spring was late and cold, lasting into June. When summer arrived, it brought heavy, drenching thunderstorms, along with vast amounts of lightning that tended to impress and terrify the citizens. Great flood damage was reported in the Severn Valley of the U.K. and in South Wales.

Krakatoa dwarfs the largest above-ground thermonuclear device, a fifty-seven-megaton affair detonated by the Soviets in 1961. (Known as a continent buster, the device was scorned as senseless overkill by jittery scientists throughout the world.)

The closest known survivor of the Krakatoa blast was Captain Watson, master of the British ship *Charles Bal.* The explosion whipped away most of the vessel's superstructure, masts, and rigging; blistered its paint; and burned anything flammable lying exposed on the deck. Vessels of this era were still a curious combination of steam drive and sail power. Captain Watson remarked that at his position—estimated to be fifteen miles from Krakatoa—the concussion was deafening.

Another captain, Joshua Stone, was farther out at sea on the *Northam Castle.* The black, murky, overcast sky and showers of mud rain prompted him to enter in his log the notation, "Surely the Day of Judgment has come." He also reported extensive damage to his vessel, as well as casualties among the crew, half of whom suffered from shattered eardrums. His vessel ended up bent and blistered, but still afloat.

The noise from Krakatoa's blast was thought to be

the loudest ever made on earth. "Ear witnesses" clearly heard it three thousand miles away in the city of Rodriguez on the British island of Mauritius, which is off the east coast of Africa, almost all the way to Madagascar.

Prior to Krakatoa, volcanoes had erupted occasionally, many with a roaring that could be heard for miles, and other volcanoes have erupted since. However, these blasts, while catastrophic to those near at hand, did not actually constitute large detonations—at least not when compared with the resoluteness, awfulness, and fury of Krakatoa.

Originally, Krakatoa was a six-square-mile series of built-up volcanic rock islands lying in the Straits of Sundra between Sumatra and Java, Indonesia. Two submerged volcanoes formed one massive 2,600-foot crater called Rakats; four additional mounts of volcanic cinder comprised the group known as Krakatoa. No one lived on the islands, but prior to 1882, they were popular outlying destinations for tourists from Djakarta. However, as group after group reported increasingly hostile conditions, visits dropped to virtually nothing. The ground was hot enough to scorch the soles of the shoes of anyone sufficiently brave to hike over the cinder cones.

An earthquake in 1880 apparently opened fissures in the earth, setting the stage for the final act. Scientists speculate that cool water under extremely high pressure was forced through the newly opened fissures and deep into the earth onto white hot molten rock. As if to say, "I won't be outdone by man's puny efforts," Mother Nature set the clock ticking on the most explosive situation in history.

Krakatoa practiced a bit before the main event. The first blast occurred at 5:30 A.M. on August 27. Dawn had just arrived. Although noteworthy, that blast was nothing more than Krakatoa clearing its pipes. Mount

Perboewatan subsided into a mighty boiling cauldron on the sea floor. So far, so good. Residents of Djakarta, 100 miles to the south, had already learned to endure the thick, ashy cloud cover and the rain, mud, and ominous rumblings.

But at 6:04 A.M., as fissures opened deeper and deeper within the bowels of the earth, Mount Danan blew up. The 1,496-foot mountain was wiped off the map. That was number two.

At 10:02 A.M., a deadly hush fell over the region. It was showtime. The hush predicted for a few precious seconds the mightiest blast ever to occur on earth.

Five cubic miles of rock disappeared in one violent shudder that literally shook the entire earth. Newly installed seismographs around the world signaled the fact that something big had happened.

In Djakarta, buildings bent and rattled with the concussion. Strong cement foundations cracked like glass jars. People, doors, windows, and any loose objects were tossed around like trash in the wind. Windows in Macassan, a thousand miles away in the Celebes Islands, were shattered. At Karim on the island of Java, 400 miles away, residents supposed a large steam boiler on an ocean-going vessel had blown up. Boats were launched in an attempt to rescue and salvage. Prisoners in Borneo 1,200 miles to the southeast seized the opportunity to riot, believing the noise heralded the onset of an attack on the harbor, giving them an excellent chance to escape. Troops in Acheen, 1,100 miles away, went on full alert, assuming they had heard a cannon attack on a nearby fortress.

But the real damage to those within hearing, but outside the impact zone, lay ahead. Monstrous killer waves were starting to ripple through the region. Twenty minutes after the big blast, tsunami waves 175 feet high traveling at 400 miles an hour hit Djakarta. Mighty walls of water washed miles inland, carrying off people and goods with impunity.

Nine hours after Djakarta, huge waves hit the harbor of Calcutta on the Indian subcontinent. Calcutta harbor was almost destroyed. Four hundred miles in the opposite direction, the tsunami tore out the new, modern harbor at Perth, Australia.

Around the Indonesian archipelago, an estimated three hundred villages were flushed off the face of the earth. Although only a few people were killed in the initial blast, the massive waves that now rippled about the earth would eventually kill thirty-seven thousand people. All ten thousand residents of Telok Betong—a city more than fifty miles from Krakatoa—were wiped out instantaneously.

A French scientific mission working in the region found evidence of tsunami damage two hundred feet uphill above some unsuspecting coastal villages. The Dutch coastal steamer *Berouw* was carried inland two full miles and dumped on a hill thirty-five feet above the coastal plane. Reverend Philip Neale, a British chaplain, reported that three thousand bodies were recovered at Merak. He located three survivors, but had no valid estimate of the total number killed. On the small island of Sebesi, another three thousand or so souls were buried under two hundred feet of water. In most cases, no census figures existed. No one knows how many people lived in these places.

Chief Inspector Abel, an engineer working above the city of Tjiringine on telegraph lines, heard the explosion and later saw the wave approaching. Looking seaward, he stood in absolute horror as the huge mountain of water hurtled toward him. Instinctively, he and his small crew raced uphill to safety. A supply cart standing one hundred feet away on the hill was tossed and broken by the giant wave. Multiton six-wheel iron railway cars were washed uphill off their tracks. Later, once the tracks were repaired, hundreds of laborers were required to work them back into place.

Official estimates compiled at the time stated that 6,500 ships and boats—from the Indian subcontinent to Australia—were lost, all broken by the giant waves. Closer to land, in the small harbors that once supported thriving villages, dead, decaying animals polluted the water. Days and night melded into each other on the islands of Java and Sumatra, the result of the extremely dense cover of ash and dust. Mud rained down out of the clouds, clogging roads and gutters and collapsing roofs.

Some of the rock blasted out by Krakatoa was light, porous pumice. Enormous pieces of it were thrown as far as fifty miles, becoming a great danger to navigation. In the Straits of Sundra, vessels coming to investigate encountered many slabs of floating pumice. The flotsam was so thick in many cases that the vessels were forced to plow through it like ice breakers. Ships' captains reported chunks eight to ten feet thick.

Because of the global impact of the explosion, and because new recording barometers and seismographs made study possible, the nations of the earth that were able dispatched scientists to Krakatoa in their best, speediest vessels.

A research vessel, the USS *Juniata*, steamed into a five-square-mile area that only two days prior had been solid, cinder-covered land. The crew took two sonar soundings in the new ocean bathtub on which they rode. One sounding was positive at 1,000 feet. At another site, they could not find bottom; finally they estimated that the sounding was about 1,100 feet in that place. Based on these observations, the estimated shot hole was thought to be 6.5 miles square by 1,000 feet thick. That much solid rock must have weighed billions and billions of tons.

Whenever the claim is made that the world has

never experienced anything similar to the explosions the large nuclear devices possessed by the superpowers are capable of unleashing, one must look back to Krakatoa. Here, we have the greatest detonation any humans have ever experienced. Other than nuclear power, there is no standard of reference. Equating Krakatoa to trainloads of TNT circling the globe seems puny to the point of being ridiculous.

Remember The *Maine*: U.S. demands vengeance!

HAVANA, CUBA, February 1898—Captain Charles Sigsbee flawlessly guided his second-class American battleship *Maine* into the tranquil tropical harbor at Havana, Cuba, the morning of January 25, 1898. Three weeks later, the vessel blew up with such fury that windows ten miles away were pulverized like dry leaves. The blast left 263 men dead, in part because the debris from the exploding vessel rained down with such intensity on surrounding lifeboats that it smashed their hulls, rendering them unusable. It was an overt act of Spanish sabotage, the papers screamed. "Remember the *Maine*. We Demand Vengeance," the headlines read.

Almost eighty years later, in 1976, Rear Admiral Hyman G. Rickover undertook the third inquiry into the cause of the destruction of the *Maine*. The first panel, in 1898, concluded that the incident was an accident. A second board of inquiry, impaneled in 1911, said the vessel was destroyed by a floating mine. Rickover's investigation relied on ancient data and testimony, photos, and technical data accumulated during the intervening eighty years. Rickover was supported in his inquiry by U.S. Navy technical specialists, naval engineers, demolition experts, and a physicist.

"We have found no technical evidence in the records examined that an external explosion initiated the destruction of the *Maine*," Rickover concluded. "Available evidence is consistent with an internal explosion alone. We therefore conclude that an internal source was the cause of the explosion. The most likely source was heat from a fire in the coal bunker adjacent to the six-inch gun reserve magazine."

In other words, the incident that precipitated a major three-year war with a large European power and cost millions of dollars and hundreds of American lives was an accident. It was not an act of sabotage or overt warfare, as the American people were led to believe in 1898.

The *Maine* was the first modern steel battleship

HAVANA, CUBA, Jan. 25—BATTLESHIP *MAINE*—In the photo on the previous page, the USS *Maine*, commanded by Captain Charles Sigsbee, sits proudly in Havana Harbor. The *Maine*, the first modern steel battleship built by the U.S. Navy, was 319 feet long and 57 feet wide. (Photo courtesy of U.S. Naval Institute, Annapolis, MD.)

built by the U.S. Navy as it emerged from the era of seven-knot-per-hour, sail-propelled vessels to seventeen-knot, steam-driven battlewagons. Her keel was laid in the Brooklyn Navy Yard in 1888.

Originally, her designers had planned for a canvas, coal-powered, armored cruiser. Such an unruly, convoluted hermaphrodite is difficult to comprehend today. Six years elapsed between *Maine*'s inception and her maiden voyage out of New York Harbor. During the intervening years, technology and military evolution overtook the vessel's designers in a manner that would become very familiar to modern Americans. The era of wind, wood vessels, and canvas sails passed irrevocably into history. In its place were steel and steam. Plans for the *Maine* were drawn, redrawn, and redrawn again.

At 10:00 A.M. on Tuesday, November 5, 1895, she stood out to sea for the first time. Old-timers were shocked to see twin stacks rather than twin masts. As a result of massive surgery, the hermaphrodite had evolved into a reasonably beautiful lady.

Maine's overall length was 319 feet, and she was 57 feet wide. Total displacement was said to be exactly 6,683 tons. Armament consisted of four ten-inch guns, two forward and two aft-mounted in steel turrets. The turrets were steam-driven by a portion of the nine thousand horsepower coal-fired engines. The fact that the boilers had to be kept at full pressure in Havana Harbor so that the guns could be deployed instantly may have led to the *Maine*'s eventual destruction.

In addition to the ten-inch guns, the *Maine* had six six-inch guns and several rapid-fire six pounders. Use of the anachronism "pounder" dated straight back to the days of muzzle-loading cannons. The *Maine* was one of the last and largest vessels in the U.S. Navy to be designated as such. Numerous torpedoes and four launching tubes rounded out the armament of the *Maine*.

In spite of what looked like a formidable amount of firepower, the *Maine* was rated officially as a second-class battleship. Four larger, more powerful vessels carrying bigger guns with greater range had been built subsequent to the *Maine*, pushing her back to second-class status. (Technology moved very rapidly during that period.) First or second class, the vessel was nevertheless extremely intimidating.

Two years before the turn of the century, Americans were aware of and involved in the affairs of Cuba. The United States had many millions of dollars invested there. Protecting these investments around the world often led Americans to actions they would not have taken otherwise. In retrospect, the U.S. War Department's attempts to intimidate Spain's colonists led to the *Maine*'s demise.

Spain began its three centuries of rule over Cuba as a colony by hunting down and killing every original native Cuban. They maintained control by ruthlessly suppressing any dissident activity. Those who settled the land were of Spanish extraction, but this did not prevent them from rising in rebellion on a fairly regular schedule.

Freedom-loving Americans tended to sympathize with the rebels, and often extended more than just moral encouragement. Guns and ammunition flowed freely from Miami into Cuba.

Quite naturally, Spanish authorities took great exception to American meddling in what they considered to be their own colonial affairs. Bloody reprisals against Cuban rebels made good material for the American press, but the issue remained, the Spanish said, outside the prerogatives of the "Yanqui Pigs." When riots in Havana threatened American sensibilities—and assets—President McKinley ordered the *Maine* to Havana on a "show the flag" mission.

In October 1887, Captain Charles D. Sigsbee was

ordered to Port Royal, South Carolina, to stand by in case of additional trouble in Havana. Two months later, as tensions continued to mount, he completely resupplied his vessel with coal and munitions. From South Carolina, he steamed a step closer, to the U.S. naval base in Key West, Florida.

Just five hours from Havana, he wrote his wife, "In certain events the *Maine* is to be the chosen of the flock, it being so ordered by the Department [of the Navy]."

Although Sigsbee left Key West in the evening, he lollygagged off Havana Harbor till full daylight when everyone was up and alert to what was happening. Large American battlewagons are truly impressive as they move into harbor with colors flying from their peaks and their white-suited seamen standing at attention, spaced evenly every ten yards along the deck. The psychological impact was as good or better than Fitzhugh Lee, the American consul in Havana, had hoped for when he asked the president for a show-the-flag battleship.

The *Maine* joined the Spanish cruiser *Alfonso XII* in the harbor, anchoring perhaps four hundred yards away. Obligatory seventeen-gun salutes were exchanged in keeping with rigid Navy protocol established years previous. The Spanish, for their part, did not greet the American visitor with any particular warmth. They did, however, display great and elaborate old-world civility.

In the weeks that followed, Sigsbee and Lee attended a number of lavish state dinners and formal receptions and were even given choice box-seat tickets to the Sunday bullfights. The riots and insurrections that had plagued Havana seemed to be a thing of the past. No one seemed to be interested—in word or deed—in pushing the giant to the north.

The day of Tuesday, February 15, 1898, was

uneventful for the two men, with not so much as a visit to a museum or a state dinner on the agenda. Weather in Havana that time of year is usually mild. Sigsbee had no reason to suspect that he was on a collision course with history or that anything was seriously amiss. Nevertheless, he kept his boilers fired so that operation of his big guns would be instantaneous. Rather than maintaining a skeleton anchor watch, he prudently ordered a full-quarter watch. When the bugler sounded taps at 21:00 hours, everything was secured.

Along with the *Maine*, the United States had ordered the light cruiser *Montgomery* to Havana as an escort for the battleship. The *Montgomery* lay gently at anchor some hundred yards to the *Maine*'s starboard beyond the *Alfonso*. There were also British, Dutch, and Belgian merchant steamers anchored in the harbor on that beautiful starlit night.

Exactly how many tons of explosive and coal dust lay in the various magazines and holds of the *Maine* seems lost in history. Given the ammo for the ten-inch, six-inch, and six-pound guns that must have been on board, the total amount can be estimated in the hundreds of tons. With a full load of coal as well, the potential for a huge multiton explosion was certainly present.

Captain Sigsbee sat at this desk, calmly writing his wife, when a tremendous blast literally rocked his vessel. Observers said the boat lifted out of the water and belched smoke at every opening, many of which were newly created by the explosion. Smoke was so thick in the corridors that Sigsbee actually had to fight his way out of the cabin to the vessel's bridge. Men carrying axes and hoses were running everywhere through the thick, acrid fumes. Had it not been for some unnamed marine who happened on him, Sigsbee never would have come to the bridge alive.

At the bridge, Sigsbee knew immediately that his vessel was doomed. He had his bugler sound the abandon-ship signal.

American newspaper stringer Walter S. Meriwether and his interpreter, Felipe Ruiz, were entering the lobby of Hotel Inglaterra when the ship blew up. Right at the door, they felt a terrific wind shock. The building and the entire city seemed to react to the deep, throaty explosion. People in other buildings were showered with falling plaster. Every light in the city went out.

"Insurgents have blown up the palace!" voices cried in the dark in terrified Cuban Spanish. Meriwether himself thought that perhaps insurgents had hit the arsenal at Regia across the harbor. He ran up to his top-floor room, from which vantage point he could see out into the harbor. There was nothing, he reported, but a great glare, and rockets seemed to be shooting up from one of the vessels.

Another American in Havana that night, Clara Barton, wrote, "The city had grown still, when suddenly the table shook under our hands. The big french windows and doors to the veranda facing the sea suddenly flew open. The deafening roar was such as perhaps one never heard before. Off in the bay, the air was filled with blazing light, and this in turn filled the sky with black specks like huge specters flying in all directions."

Across the harbor, British Captain Frederick G. Teasdale thought for a moment that someone had shot him in the head, the concussion was so great. He ran from his cabin out onto the deck of his bark (a three-masted schooner), where he saw a great cloud of debris hurling high into the sky over the harbor.

Sigmund Rothschild, a passenger on the American vacation cruise ship *City of Washington* was looking right at the *Maine* from about six hundred yards when she went. He saw the bow rise out of the water, felt the

HAVANA, CUBA, Feb. 15—BLAST ROCKS VESSEL—The great battleship USS *Maine* after a huge explosion blew it out of the water. Within five minutes of the blast, the ship was resting on the harbor bottom thirty feet below the surface. (Photo courtesy of U.S. Naval Institute, Annapolis, MD.)

ARLINGTON, VA, Mar. 23—REMEMBER THE MAINE—U.S. servicemen recovered from the wreckage await burial at Arlington National Cemetery. The explosion of the USS *Maine* left 263 dead. (Photo courtesy of U.S. Naval Institute, Annapolis, MD.)

HAVANA, CUBA, Mar. 30—SUNKEN SHIP AWAITS SALVAGE OPERATIONS—Twisted remains of the USS *Maine* (below) await salvage operations after an explosion that was originally thought to be an overt act of Spanish sabotage. (Photo courtesy of U.S. Naval Institute, Annapolis, MD.)

great concussion, heard the deafening roar, and then heard the noise of junk falling onto his own vessel and into the surrounding ocean.

Five minutes after the explosion, the *Maine* was resting on the harbor bottom thirty feet below. Undamaged lifeboats from the *City of Washington* and the *Alfonso XII* quickly arrived to pick up a small, pitiful band of survivors. Spanish seamen reportedly spared no effort to help the Americans, despite the ever-present danger of being blown to hell themselves. On shore, Spanish doctors and nurses similarly spared no effort to treat the wounded. Some were taken to San Ambrosio Hospital. Others went to the Havana City Hospital. Reporter Meriwether watched in despair at what seemed an endless line of victims being brought in on stretchers.

The news regarding the destruction of the *Maine* spread rapidly. Underwater telegraph cables carried the first news accounts to Miami in about twenty minutes. It was 1:30 the next morning when the news reached the Secretary of the Navy in Washington, D.C.

Chief of the Bureau of the Navy Francis Dickens immediately walked over to the White House, where he wakened President McKinley. The message from Cuba urged that "public opinion should be suspended until further proof of the disaster's cause is established." (Perhaps a dozen or more fires broke out spontaneously and accidentally on coal-fired fighting vessels during that era. Because these vessels were not fully prepared for battle, with bulging magazines containing thousands of rounds of high explosives, the results were not devastating as in the case of the *Maine*.)

Contemporary findings supported by the cool, steady objectivity of history support the conclusion that the incident was, in fact, an accident. At the time, however, *New York Journal* owner William Randolph

Hearst, when told of the incident, asked his city editor, "Have you put anything else on the front page?"

"Only the other big news," the editor replied.

"There is no other big news," Hearst screamed. "Spread the story all over the page! THIS MEANS WAR."

And so it was. America went to war with the cry, "Remember the *Maine*!"

Mysterious projectile detonates in mid-descent; flattens 800 square miles in Siberia

TUNGUSKA, SIBERIA, June 1908—On December 11, 1903, the Wright brothers made the world's first manned airplane flight. It is therefore extremely unlikely that remote, uneducated Siberians had any such frame of reference four and one-half years later when a "streaking silver object" crashed and detonated in one of the most obscure, inhospitable regions of Russia. The region was so remote that it wasn't until 1928, fully twenty years later, that a research party reached the site of the explosion for a firsthand look.

Although the evidence is extremely sparse, gathered entirely by inference and conjecture, the blast may have been among the largest ever unleashed on earth. Willard F. Libby, a Nobel prize-winning chemist, calculated that it was roughly 1,500 times the size of Hiroshima, or the equivalent of thirty megatons of TNT. Nobody really knows what produced the great blow to Siberia or knows with any certainty how it happened. If Tunguska did, in fact, rival known blasts such as Krakatoa, it cannot be proven.

Early in the morning of June 30, 1908, witnesses on and around the Indian Ocean reported seeing a shining, silvery, cylindric object plummeting from high in the heavens in a long, shallow trajectory. It continued its descent over the Gobi Desert of China, where herdsmen and caravan drivers reported the strange, unfamiliar object. By this time, the projectile, on its descent into earth's Northern Hemisphere, had reached the atmosphere. It began to glow brilliantly.

Strongly attracted now by the earth's gravity, the object continued on into the most remote area on earth, impacting the gently rolling Siberian scrubland 590 miles north of Irkutsk, U.S.S.R. Two decades later, researchers identified the exact spot, fifteen miles north of the Stony Tunguska River.

Scientists are still uncertain about the projectile, but whatever it was, it detonated with extreme violence about three miles above ground level. The fact that the object—meteor, spaceship, nuclear bomb, or whatever—did not dig a crater added immeasurably to the mystery. When researchers finally reached the impact zone in 1928, they expected to discover large quantities of nickel-iron meteor fragments imbedded in the taiga or Siberian permafrost-bound swamps. But there was nothing except severely scorched trees and tundra, preserved and left intact by the oppressive cold.

Mr. S.B. Semenov, a farmer in the village of Vanavara, forty miles southeast of the blast, was the nearest living witness that could be identified. He reported that while sitting on his porch, he suddenly saw an enormous fireball covering most of the sky. The heat radiating from it was intense enough to scorch his shirt and burn exposed skin. Closely following the amazing fireball was an immense explosion that threw farmer Semenov right out of his chair and off the porch.

To some extent, superstitious Siberians made the job of collecting precise information very difficult. Most who had seen the event or were subsequently in the area were extremely reluctant to come forward because of the almost religious mystique that developed around the incident. Research was especially difficult twenty years later, after superstitions regarding the incident had spread and strengthened.

Llya Potapovich, another farmer/herdsman living in a remote station named Tayshet, finally reported that his now-deceased family member, Vasily Ilich, had kept his herd of 1,500 reindeer in the exact blast zone. Ilich had also owned a number of storehouses containing food, supplies, clothing, and harnesses. Potapovich said Illich told his family that all were incinerated beyond recognition.

On June 30, 1908, passengers on the Trans-Siberian Railway, 375 miles to the south, reported hearing three distinct blasts and experiencing severe shock waves. Concussion from the distant blast was sufficient to cause the engineer to bring the train to an immediate halt. People as far as 550 miles away reported that they heard the blast plainly. In Irkutsk, the tremor was recorded as a significant earthquake. Similar seismographic readings were made in St. Petersburg (now Leningrad), London, New York, and Jena, East Germany, among others. This was a very curious circumstance, since all on-site observers eventually agreed that whatever it was never touched the earth directly.

In Siberia, the intense heat was strong enough to melt the permafrost, causing severe flooding. Water in the Stony Tunguska River flowed over the banks for several years after the blast. Following the incident, a high-altitude silvery cloud of a strange luminescent quality blanketed Siberia for almost a week. It reportedly produced light so intense that people could see plainly throughout the night. Extraordinary dust clouds, along with strange, unexplained nocturnal displays, were evident for several weeks in places as distant as Vienna, Copenhagen, London, and Antwerp.

As a point of contrast, the Hiroshima blast devastated an area of about eighteen square miles. The Japanese blast torched flammable material as far as four miles away. At Tunguska, nearly eight hundred square miles were flattened. Wood ten miles from the blast center caught fire. Dozens of small, disk-like holes gave the impression that fast-moving, smaller objects ricocheted off the marshy taiga in the immediate blast zone.

Closer in to the suspected ground zero, scientists were puzzled to discover great areas of birch and aspen trees stripped of leaves and branches, standing like a forest of telegraph poles. They were at a complete loss to explain the peculiar pattern of these trees, surrounded by millions of trees further away from the blast zone that were blown over at their bases. More curiously, the trees lay windrowed into a wave-like pattern. The incredibly cold, inhospitable Siberian climate had preserved most of the scars of the blast until the first outsiders arrived at the site twenty years after the blast. Photos taken at that time look as though the fire had been extinguished the week previous.

Nomadic Mongol natives of the region reported thick, sooty, black rain accompanied by dense clouds and continuous thunder for weeks after the blast. As an added mystery, the falling ash deposits caused a strange, scabby rash to appear on the backs of their horses and reindeer. Like the scientists who studied the Tunguska phenomena, natives who had actually seen it offered no valid explanation as to what had happened.

As time went by, roads were pushed into the region by the central government. Eventually, the government even put in a short landing strip so that the area could be surveyed by air for signs of who or what. Great numbers of theories were advanced to explain the situation. Some excellent scientists joined the ranks of the occult, suggesting that the blast was nuclear in nature and postulating in most convincing terms that a spaceship had actually visited the earth in 1908 and blown up.

By the late 1940s, when testing was first done for radioactivity, all traces had faded from the scene. However, Siberian scientist Aleksandr Kazantev, who visited Hiroshima as an official spectator for the Soviet Union, saw great similarity between the ground effect of the Tunguska blast and that of Hiroshima.

When detonated, the great heat generated by nuclear weapons approaches five thousand degrees Celsius. Everything within eight to ten miles is set on fire. Shortly after the blast, the superconcussion generally snuffs out all of the fires. As part of the third stage, a huge amount of debris is sucked high into the atmosphere, where it roils and boils, creating severe thunderstorms. All of these conditions were present on the ground and in the air at Tunguska, as described by observers.

Tree rings formed during and after the period of the blast suggest high levels of radiation were present. There is also some indication that other regional plant life was genetically altered. No radioactivity registers on instruments today, but neither does it in Hiroshima or Nagasaki.

Some scientists believe it was a meteorite that heated and disintegrated at an altitude of fifteen thousand feet. If so, it must have weighed at least four thousand tons. Meteorites of that size visit the earth on a very infrequent schedule.

Curiously, it appears that the flight path of the

object may have changed dramatically—almost as if something were steering it—after it entered earth's atmosphere over Siberia. Patterns of destruction on the ground have led some Soviet scientists to conclude that the object circumscribed a huge, 375-mile-long arc, something they insist no naturally falling object could accomplish.

Whatever it was created a blast so huge that the only possible historic comparison is Krakatoa. Seismographic recording instruments in London and New York were impacted twice as the concussion circled the earth and hit the sensitive gauges a second time. Observers at the time did not know what they were dealing with. In many instances, the readings were not recognized for what they actually were for thirty to forty years.

Eventually, Soviet researchers located people—scattered over an area easily the size of France and Germany combined—who saw or heard the blast.

Because the region was populated by nomadic Mongols who remained basically unknown to central authorities, no one knows for sure whether there were human casualties. There may have been four or five people who perished in the blast. The very existence of these hardy, reclusive Siberians would have been known to perhaps a dozen others. Scientists believe that whatever it was could only have produced a few casualties, having come to earth as it did, in the midst of what made up a maximum of five percent of the earth's surface. (Had it landed anywhere on the other 95 percent, casualties would have been far greater. This, many claim, is further proof that intelligent beings were attempting to steer some sort of craft away from places where it would do considerable damage.)

Whether one leans toward science fiction or toward pure science when evaluating the Tunguska incident is a matter of personal choice. We will probably never know the cause of the blast with absolute certainty. But evaluating the observable effects of this incident seventy years later easily leads to the conclusion that it was one of the world's major explosions.

53 tons of TNT detonate at Silvertown purifications plant causing widespread destruction

SILVERTOWN, ENGLAND, January 1917—English historians commenting on the great blast in 1917 in Silvertown, London, claimed it was the largest explosion to date. The claim would have been valid if one ignored Krakatoa, the battleship *Maine,* and Tunguska. Perhaps these folks were referring only to man-made explosions within very large cities. In that regard, the blast was definitely world class. Most countries insist that explosive manufacture be conducted in remote, sparsely populated regions.

In addition to being one of the world's outstanding free-ranging explosions, Silvertown also included elements of sabotage and clandestine activity not found in any competing blasts. Records regarding details of the blast and reports from a board of inquiry's investigation into the detonation were sealed in 1917 for a statutory forty years, so as "not to furnish information to the enemy." Even in 1957, when researchers attempted to look into the matter, obtaining any official reports proved to be exceedingly difficult. It wasn't until late in 1974 that a "secret" report was located and released, allowing any sort of valid explanation of the incident to be set forth.

In the tense wartime atmosphere of 1917, rumors circulated locally that German agents trained in America had destroyed the TNT plant near Crescent Wharf, or that invisible German zeppelins that produced their own fuel as they moved through the sky had bombed the factory in honor of the Kaiser's birthday.

Perhaps these were typical of wartime rumors, but in retrospect they seem ludicrous. Obtuse references to Americans indirectly causing the blast are especially curious. Perhaps the British were suffering horribly and were upset that Americans did not take part earlier and with more enthusiasm.

As it is, there is reasonable evidence that the plant could have been the victim of secret agents. If these theories are true, Silvertown is one of the only great detonations to be induced intentionally as an act of

LONDON, ENGLAND, Jan. 19—BLAST FLATTENS SECTION OF SILVERTOWN—The photo on page 27 shows how the huge explosion destroyed every building within four hundred yards of the Silvertown TNT purification works. Damage to the industrial/residential community was substantial as far out as 1,500 yards. (Photo courtesy of Imperial War Museum, London, England.)

war in the midst of a heavily populated area.

Several years into World War I, English military planners found they were heavily constrained by a lack of high explosives. Consumption at the front had occurred at such a wildly enthusiastic rate that production failed to keep pace. Destroyers were dropping thousand-pound depth charges on submarines in clusters of six or eight. Even new airplanes were part of the problem as they promiscuously released charges of twenty-five to fifty pounds. All of this created a previously unappreciated demand for high explosives.

As a result, the Ministry of Munitions made an inquiry into bottlenecks in explosive production. Government experts determined that the problem lay in the industry's inability to "purify" TNT as rapidly as it could be manufactured. While explosive manufacture was relatively straightforward, taking out impurities that could and would cause premature detonation was another matter.

Recognizing that the purification process was far more hazardous than the initial manufacture, the British government set up an incentive program paying five shillings per ton to companies that would take up this enterprise.

Brunner Mond & Co., surviving today as Mond Division of Imperial Chemical Industries, was already heavily committed to the primary manufacture of explosives for the military. Because such enterprise was entirely underwritten by the British government, getting into the secondary business of purification consisted of an exercise in allocating scarce wartime resources rather than a financial risk. Virtually all English factories were running at full capacity. Unusual "open" manufacturing facilities were about as scarce as German teachers in British high schools. Brunner Mond's challenge was to locate a facility, not to figure out how to pay for it.

After a diligent and somewhat protracted search, the company found an unused facility in the Silvertown region of London at a place called Crescent Wharf. The factory had previously been used to handle overflow production of caustic soda. It was flanked on the west by a plywood and packing case manufacturer and on the south by the Silvertown Lubricants Company. Some three thousand people lived within a quarter mile of the proposed purification plant. At best, movement of the explosives in and out was an inordinate risk. Company officials later said they chose the site because it already contained much of the machinery they needed to get started.

Even the government later admitted that as a place to handle high explosives, the dingy waterfront location was far from ideal. Yet, because of the scarcity of open factories and the fact that most of the needed machinery was already in place, the urgency of warfare won out. Anxious company and government officials opted to overlook the risk to the local population. Had they decided to build or to go elsewhere, it is doubtful that the factory would have been brought on line before the end of the war.

Caustic soda—the factory's formerly manufactured product—and raw TNT definitely are not compatible. The combination of destructive materials could easily cause premature detonations. Cleanup work was said to be thorough and complete, but it was done under intense pressure to get the plant on line. Initial output from the plant reached the front early in 1916.

The degree to which the TNT was purified greatly affected its stability and shelf life, as well as its total destructive force. Even as late as World War II, most nations continued to experience problems with high-explosive ordnance that exuded dangerously out of its casing after only brief storage. Purified TNT tended to remain stable in its crystalline state when packed in high-explosive rounds.

At the time, the process of TNT purification consisted of emptying fifty-pound cloth bags packed with raw TNT into a five-ton melt pot, which was warmed by a closed-system hot water jacket. Fire for heating was kept away from the melt pot; only the hot-water circulated to the explosives.

After being warmed, the crude TNT was dissolved in an alcohol bath. Upon redrying and cooling, the TNT crystallized. This crystallization process partially purified the material; remaining impurities were removed by centrifuging out residual alcohol. Completely dry, crystallized TNT was dug out of the vat and repacked in special fifty-pound cotton bags. Any remaining uncrystallized material was returned to the pot for another run through the process. Theoretically, some material was recycled three or four times before being pulled out as purified TNT. All alcohol used was carefully collected, filtered, distilled, and reused. The dark brown residue from the process, aptly named TNT oil, was soaked in diatomaceous earth to produce nonfreezing commercial dynamite. This by-product alone—constituting approximately four percent of the original crude TNT that came into the factory—was sufficient to make the whole project economically viable.

Sixty-three people were employed at the Silvertown TNT works. Afternoon shift workers started in at 2:00 P.M. One fortunate laborer was absent, but the other twenty on that fateful Friday-afternoon shift were at their appointed stations at the appointed time. Because of the nature of the process, the factory ran 'round the clock. Aside from a few men required to accomplish especially rigorous physical tasks, girls and women ran the factory's day-to-day operations. Most of the men, of course, were in France, dying in the trenches.

At 6:40 P.M. on Friday, January 19, employee Hetty Sands decided she should climb the long wooden staircase up to the top of the melt-pot room containing the heavy bags of crude TNT. The bags themselves were moved via a small electric freight elevator. By squeezing, she might have used the elevator rather than walking, but Sands wanted to "pop up" and then down again without creating any fuss. She was a diligent worker, but, like many in the U.K., had grown tired of the dull, back-breaking work women like herself were forced into by the war. Her morale might have sagged were it not for the prospect that she could quit as soon as the war did.

Up at the melt-pot room, she noted that Walter Mauger was taking the final three bags from the small elevator hoist, which was securely fastened to its moorings at the floor of the melt-pot room. Using a small clasp knife, Catherine Hodge was cutting the bag strings and emptying the contents into a small chute leading to the cast-iron melt pot. Everything looked to be in order.

Mauger told Sands that they now had enough material to start the next melt batch. He told her she could go on break without disturbing the factory's regular flow. She joined Ada Randall at the bottom of the stairs, and together they started towards the women's lavatory.

By modern standards, the women's facility was placed at much too great a distance from the work area. At the loo, the two women heard two faint thumps, much like a heavy door banging shut. They ran out just in time to see the top of the melt-pot room enveloped in bright orange flames. Desperately, they ran towards the factory's perimeter fence. It was the shortest route from the loo away from the developing holocaust, but they later admitted they had no idea how they were going to cross to safety. There were no gates or openings at that particular point on the fence.

Crossing the fence was a problem that the women did not ultimately have to address. At 6:52 P.M., fifty-three tons of TNT resolutely detonated in the closed industrial/residential community. The fence—150 yards from the melt pot—was blown flat, and so were the women. Most of their clothing was carried away. After picking themselves up, however, they were able to proceed through the rubble to safety.

The sound of the blast was heard plainly fifty miles away in Cambridge. Every building within 400 yards was destroyed. (Although there were many buildings, most of them were ultimately hauled away.) Out to about 650 yards, damage was substantial. As far out as 1,500 yards, windows broke, roofs whooshed away, and weaker wooden sections caved in.

St. Barnabas Church and the meeting hall across North Woolwich Road were leveled. A fifteen-ton boiler was thrown like a child's toy into the factory across the street.

In those times, fire department personnel lived in their fire stations. The crew at the nearest station, conveniently located kitty-corner across the street, had just begun to roll out their trucks when the blast hit. Two of the firemen were killed instantly, and a third woke up in London Central Hospital the next day. Along with the firemen, three members of their immediate families also became casualties. Additional fire brigades from central London eventually responded to the emergency, but the blast disrupted communications, causing what many felt were undue delays. Fires raged through the area throughout Friday night and late into Saturday. In that era, methods of dealing with oil fires such as the one at Silvertown Lubricants were only poorly understood. In an attempt to gain the upper hand over the flour-mill fires, officials floated in a pump barge with tremendous capacity to lay water on the blaze.

Of the nine men who showed up for the 2:00 P.M. shift, only a boy named James Arneli, who worked in the centrifuge, survived. Women laborers, of which there were eleven, were more fortunate. Only a female employee who worked in the melt-pot room died. The ten others survived to tell their stories to the board of inquiry examiners.

Board members determined that the unattended bags of crude TNT that traveled from the primary manufacturing site first by rail, then by barge, and at last arrived at the plant by truck, could have been tampered with. A stick of caustic material quickly inserted into a bag could, when it reached eighty degrees Fahrenheit during the melting phase, have caused the bag to burst into flame. Small amounts of TNT, as has been found time after time, will burn without incident. Yet more than a pound or two will burn long enough to heat sufficiently to detonate.

Enemy agents could have inserted a stick of caustic soda wrapped in linen cloth into a random bag at any place along the material's journey. At the purification plant, the bags were simply dumped uninspected into the melt-pot hopper. Investigators were sufficiently concerned about the Silvertown blast that they locked the records away for fifty-seven years. We will never know the cause with certainty. Whatever it was, it resulted in one of the grandest inner-city explosions of all time.

British detonate explosive-packed maze of tunnels below German lines

MESSINES RIDGE, BELGIUM, June 1917—Strange as it may be, the vast majority of the world's great explosions have been nonmilitary in nature. There have been a few exceptions that were *closely associated* with military undertakings. But Messines Ridge was the only one that was purely military. It was also a grand blast that can take its place with the other notable detonations that have come down to us through history.

At the start of 1915, Europe was the scene of two fairly evenly matched contenders whose opposing forces were stalemated in dreadful, muddy, wet trenches. Because poor communications and transportation technology severely restricted both sides, neither could exploit, much less engineer, a breakthrough. Try as they might, neither side could secure an advantage, either in equipment, command, or tactics.

Life in the trenches for the soldiers of both sides

cannot be characterized easily, it was so horrible. Great seas of mud covered the lines. Men's feet literally rotted in their boots. Insignificant scratches from the concertina wire turned into large, weeping, gangrenous wounds. Illness and plague were endemic. Entire forests were doomed by constant artillery fire. Every square yard was in someone's line of fire.

Death in the grimy, narrow, water-filled trenches was often more certain—and more grisly—than life. The lucky ones died quickly, compressed to raw protoplasm in the gooey mud. Photos taken after gas attacks show hundreds of the living cowering among thousands of the dead, heaped together in the narrow trenches.

About the best a common enlisted man could hope for was an incredibly miserable six weeks of standing, sleeping, and eating in syrupy mud, only to have his heart shot out by enemy machine gunners when ordered over the top, out of the shelter of his earthen mausoleum.

Losses to both sides were occurring at a rate in excess of 5,500 men per day in 1915. In England, as old generals fought yesterday's wars against today's technology, the losses were so severe that there was talk of not pursuing a serious offensive in 1917. Costs in terms of lives, vis-à-vis territory gained, made any kind of offensive very unpopular. German attrition by disease or accident could, the generals admitted, tip the balance in the Allies' favor. Other events intervened, however.

April 16, 1917, saw the French summer offensive at Aisne disintegrate after dozens of divisions were sacrificed for nothing more than a four-mile advance on a sixteen-mile front. To reach this level of casualties, French generals had packed an entire division together behind every thousand yards of front. French soldiers were on the verge of mutiny over the senseless loss of

200,000 men for such small gain. Without additional aid, it seemed possible that the French might throw in the towel.

In London, First Sea Lord Sir John Jellicoe dropped the other shoe when he announced that time was not on the side of the Allies and that he did not feel they could wait for attrition to weaken the enemy. German submarines were taking such a toll on merchant shipping that it would be impossible for Great Britain to continue the war in 1918, he said. The Royal Navy would be in grave difficulty unless the army cleared the Belgian coast. Jellicoe had been characterized by some as being "an old woman;" yet, he was the First Sea Lord for a country with perhaps the most powerful navy in the history of the world. Peers might have questioned his logic at times, but they did listen.

It was in this context that the British Commander-in-Chief, Sir Douglas Haig, decided to open a fall offensive on the Ypres in Flanders. The entire action spanned the twelve weeks from June 5 through November 10, at which point the offensive became helplessly bogged down in the unseasonable mud. As part of his strategy, Haig put Sir Herbert Plumer in charge of the British Second Army. Plumer was described by Bernard Montgomery, who served under him at the time (and subsequently went on to fame in World War II), as "a soldier's soldier, held in trust and respect by his men." He was a clever and methodical person whose plodding nature rubbed off in large measure on Montgomery. (Montgomery was later accused by U.S. General George S. Patton of trying to fight World War II using World War I trench warfare systems.) Plumer took to its highest level the science of persistently applying sheer weight in order to break the enemy front.

General Plumer actually started planning his mining operations in 1915. Eventually, this allowed his

Second Army to take Messines Ridge with relatively light losses by tunneling under enemy lines, emplacing high explosives, and blowing up the Germans and their carefully constructed defenses.

Rather than being sent to the trenches as cannon fodder, skilled Welsh miners allowed themselves to be recruited for the tunnel-building projects. Unskilled laborers, it was discovered, were not equal to the task at hand. Work progressed only as fast as there were skilled Welsh miners to manage it.

The concept of breaking through enemy defensive lines by undermining them was not new. The tactic had been tried on numerous occasions since the time explosives were first employed in warfare and soldiers took positions behind strong defensive structures. Black powder, combatants soon found, was not particularly adapted for use as a sapping charge.

With the advent of nitrated cellulose and this class of explosive's relatively fast and high detonation rates, various commanders again thought of deploying sapping charges. In theory, all one had to do was dig in under the enemy lines, enlarge a cavern at the end of the tunnel, fill it with explosives, and blow the enemy's position to hell.

The previous lack of suitable explosives was not the only reason the strategy had fallen into disrepute, however. It also had to do with the demise of castle defense. Lines no longer remained static enough for serious mining. Tunnel digging often had to be accomplished under adverse rock and soil conditions. Cave-ins and washouts were common. Often, water-filled tunnels were driven through places far from ideal for such activity. To a great extent, mining—whether for sapping activities or for commercial reasons—progressed no faster than the technology of pumps used to keep them free of water.

At Messines Ridge, the hardy Welsh miners

encountered interbed after interbed of clay on which numerous strata of water stood. As an additional hazard, the enemy listened for underground activity. When the miners were detected, the enemy either blew in the tunnels by using small charges called camoflets, or dug down and appropriated the tunnels for their own use.

Yet, working to the extent of the Welsh miners' endurance, the British were able to hack a total of twenty-four tunnels into Messines Ridge, one hundred feet below the German lines.

Messines Ridge constituted a military prominence in the region, giving whomever occupied it a decided advantage. General Plumer wrote, "Not one square yard of our front is not under enemy observation." The British were forced to muck about in the porridge-like mud surrounding the ridge, mostly under the watchful eye of the Germans. Hill 60 in the group was the key to the ridge. German commanders thought their positions were impenetrable because of the terrain and the intricate system of trenches, mortars, and machine guns defending the hill. German artillery averaged one gun for every seven yards on Messines Ridge.

More than a million pounds of an explosive called Ammonal were eventually packed into the tunnels. (Ammonal was a common, relatively easy-to-handle explosive used by the Germans, French, and English during World War I. Basically, the explosive is ammonium nitrate mixed with a small amount of TNT and aluminum flakes. As the war progressed and TNT became scarce, it was replaced in the mixture by more aluminum flakes and ammonium nitrate. Both the French and the Germans used Ammonal in major-caliber ammunition as part of the exploding charge. The flaked aluminum made it a very flashy, showy explosive.)

Bringing that quantity of high explosives up to the front and then packing it into the tunnels by hand was a difficult and harrowing chore using World War I technology. As the mountain of explosives packed beneath the enemy grew, apprehension among the British increased at a commensurate rate. Hand-powered wooden-wheel carts transported blocks of Ammonal along boarded roads built into the mines. No smoking was permitted. Feeble, flickering kerosene lanterns that provided light for the men in the tunnels were tended with almost reverent care lest they malfunction and detonate the explosive prematurely.

Although General Plumer originally conceived the sapper project in July of 1915, and some initial work was done at that time, serious progress did not begin until January of 1916. Detonation occurred on June 7, 1917—twenty-three months after conception and eighteen months after intense labor began. The time span is certainly a poignant comment on the conditions under which World War I was fought.

Part of the program consisted of a constant artillery barrage, which kept the Germans close to their holes. British aircraft patrolling the region intensely prevented enemy aircraft from spotting work on the mines.

A total of twenty-four shafts comprising about five miles of tunnels were ultimately bored through the mud and clay. Four of these tunnels missed their mark and were not used. One was discovered by the Germans and abandoned. Four of the nineteen tunnels eventually used were completed at least one year ahead of their intended use date.

Apparently, the German Commander Otto von Below was not totally trusting of the protection afforded by the waterlogged ground surrounding his position. He was alert for evidence of tunnel building. When he detected anything suspicious, he made use of his small camoflets to discourage the British. As a result, the Germans severely restricted activity in one shaft and completely closed another. A countershaft being dug by the Germans from Hill 60 was, after careful calculation, determined to be inconsequential. Although the British predicted that the German shaft would eventually reach theirs, they also figured it would not do so before the day of their Big Shot.

It rained continually on the afternoon and evening of June 6. By 2:00 A.M., the clouds broke and the sky cleared. At 3:10 A.M., the British let her rip. Haig wrote that it was the biggest explosion in the history of man. He was, of course, wrong, but the blast was, nevertheless, impressive. Prime Minister David Lloyd George knew the blast was scheduled. He stayed up all night listening for it and got his wish—plainly hearing the explosion in London, two hundred miles from its origin. Allied soldiers in northern France also reported feeling or hearing the blast.

When it went, the hill turned red and then purple with fire from the colorful explosive. Nineteen fiery volcanoes were etched against the dawning sky. Eyewitnesses said the ridge looked as though a giant had tossed it in the air.

Even German accounts of the action do not reveal estimates of casualties. They probably didn't know how many died. Several extra divisions were packed into the area in anticipation of a British offensive.

As soon as the detonations were completed, British soldiers stormed the position in relative safety. Allied artillery opened their most merciless and active barrages, attempting to limit reinforcements and resupply. This constant British shelling had kept resupply to an absolute minimum; German soldiers suffered immensely from a lack of food and water. Often they drank from filled shell holes, not knowing what human pieces lay at the bottom of the craters. But a lack of communications ability forced the British to stop short or to risk shelling their own rapidly advancing troops.

The British lost sixteen thousand men that day, gaining a total of two-and-a-half miles on a ten-mile front. These casualties were characterized as "very light." They captured a total of 7,200 German soldiers.

Messines Ridge was insignificant, according to a German commander writing at the time. He pointed out that it took the British two years to prepare the action and that it netted them a mere two and a half miles of territory. At that rate, he observed, it would be a long, long time until they would succeed in driving the Germans back to Germany.

It was, however, one of the few instances of the use of siege tactics in a siege war, and it was a very large blast.

French ship loaded with TNT blows up in Halifax Harbor, devastating city

HALIFAX, NOVA SCOTIA, December 1917—From 9:17 A.M. on Tuesday, December 6, 1917, until 5:29 A.M. on Tuesday, July 17, 1945 (when the United States detonated the world's first nuclear device in the Alamogordo Desert), the largest single explosion of human origin had occurred in the harbor at Halifax, Nova Scotia. During the intervening twenty-eight-and-a-half years, common reference has been made to the incident. Many of the tales are apocryphal, but they do tend to illustrate the extent to which the incident was etched in people's minds.

The first nuclear blast was calculated to have had an explosive force of 20,000 tons of TNT. A total of between 3,000 and 4,000 tons of TNT—depending on whose figure one reckons to be most accurate—was stored in the hold of the *Mont Blanc* at the time she went up in Halifax Harbor. At the time of the explosion, TNT (or trinitrotoluene), produced from toluene and sulfuric and nitric acids, was the most powerful explosive known to man. TNT has an effective detonating velocity of 21,000 feet per second. Ammonium nitrate, by contrast, explodes at a relatively benign 3,600 feet per second. A very fast rifle hurls its projectile at a comparatively leisurely 3,300 feet per second.

Assuming a cross-country distance of 2,500 miles as the crow flies, New York to San Francisco, a string of TNT would take approximately 10.5 minutes to detonate across the United States. A chain of ammonium nitrate, on the other hand, would detonate at a more slothlike pace; theoretically, it would require a full 61 minutes to run coast to coast.

If, as some records indicate, the total amount of TNT at Halifax Harbor was 4,000 metric tons (as opposed to 3,000), that equates to 8,800,000 pounds; even the lesser figure would have been 6,600,000 pounds worth. The 3,000-ton steel vessel vaporized instantly. Obviously, a paper manifest for what was a secret wartime cargo did not survive for our perusal,

HALIFAX, NOVA SCOTIA, Dec. 6—FROZEN WASTELAND— On page 43 is a view of the frozen wreckage that was the town of Halifax following the explosion of the French ship *Mont Blanc*, which was loaded with thousands of tons of military explosives. More than two thousand people were reported dead or missing, and property damage was estimated at over $50 million. (Photo courtesy of Imperial War Museum, London, England.)

and courts of inquiry were somewhat vague about the exact quantity of explosives (due in large measure to attempts at maintaining national security at that time).

Europe was in the grip of World War I on December 6, 1917. Canada had been in the game for several years, having declared war on Germany a few days after the British did on August 5, 1914. The United States had only recently gotten into the fray, but in trading with the Allies over the previous three years, it had suffered some shipping losses at the hands of German U-boats.

Halifax Harbor was one of the principal ports of call for supply vessels taking the Atlantic route out of Glasgow and other European ports. As a result, the region was prospering mightily in 1917. To date, no enemy vessel—either surface or underwater—had threatened the port.

Mont Blanc, an otherwise nondescript French survivor of the Atlantic killing grounds, called first in New York Harbor, where she was loaded with military explosives to be used toward the effort in Europe. (In all probability, the TNT came from DuPont, Atlas, or Hercules in Delaware. Business had been very good for companies in the United States, which had until very recently managed to avoid direct involvement in the war.)

From New York, the *Mont Blanc* made the relatively short voyage to Halifax, intent on topping off with a consignment of coal. Steep brown hills rose out of the icy gray water on either side of the Nova Scotian narrows at Halifax. Winter is depressingly bleak in that part of the world. No doubt the captain and crew reflected on this, as well as on their chances of survival if fate ever delivered them into the raging North Atlantic via one of the Kaiser's *unterwasser schiffen*. The captain had flown his obligatory red warning flag as he loaded in New York, but not as he hugged Halifax Harbor's north shore, proceeding at a cautious seven knots toward his rendezvous with fate.

Mont Blanc's crew idled their vessel off the Nova Scotia coast the night of December 5, waiting for full daylight to make the passage up the congested channel into Bedford Basin. At forty-five degrees north latitude, Mother Nature is extremely stingy with daylight that time of year. At first light—about 8:30—they turned the ship almost due west, up past McNab Island and into the narrow inlet to one of the most protected basin harbors in the New World.

Inside the narrows leading the last mile or so into Bedford Basin, the French shipper encountered a much larger Norwegian freighter loaded with coal and heading out to open seas, bound for Belgium. Traffic was heavy in the three-quarter-mile-wide channel as the many shippers attempted to take advantage of what little daylight was available. During those times, navigational aids were limited to lighthouses, marker buoys, and horns.

Captain LeMedec of the *Mont Blanc* signaled for a port passage. The Norwegian *Imo*'s skipper thought it should be a starboard passage between shore and the vessel, and he signaled accordingly without waiting to hear the *Mont Blanc*'s response. Mass confusion resulted. Observers on shore were not particularly alarmed, as fender benders among the many vessels clogging the port were common. The notoriously irritable *Imo* captain reportedly was upset over having to wait an extra day to load coal for the return to Europe; he was in a hurry, doing everything by instinct. A starboard passage might have been proper, but not under the circumstances.

Captain LeMedec, who had not, up until this time, revealed the exact nature of his cargo to his crew, swung his vessel around in desperation, but the *Imo*'s prow cut through it ahead of the main hold where the TNT was stored.

It was not a brutal crash, yet it was a most fateful

encounter. TNT is relatively insensitive to shock. It will not detonate as the result of being struck by a single rifle bullet. Sustained machine gun fire can sometimes cause an explosion. The brush might not have caused a detonation even if the explosive had been impacted.

As it were, *Imo*'s bow crushed a drum of benzol stored in the forward compartment. The fluid ran across the deck and down into the more rearward storage holds. Sparks, either from the collision or from a cooking fire, ignited the flammable liquid. (TNT is not normally considered to be hazardous in a fire. GIs—then and now—burn chunks of it to heat their coffee. Military manuals suggest that up to a pound of the material can be safely ignited without fear of detonation. However, at least one GI has cut too large a plug with which to heat his coffee. When he hurriedly tried to stamp the flame out, the explosion tore off his heavy boot.) On-shore observers reported intermittent hot, blue flames preceding the actual detonation by several minutes. Apparently, the explosives tried to burn away quietly, but the critical mass was too great.

On being told the nature of the *Mont Blanc*'s cargo, the crew quickly and silently abandoned ship, desperately pulling oars toward shore with all their might. Citizens on shore, alerted to some strange, unfathomable drama by the *Imo*'s blaring horn, watched as the vessel backed off several thousand yards and beached itself. There is no indication that its master knew anything about the delicate cargo in *Mont Blanc*'s main hold; he apparently beached for fear of sprung plates.

Some of the men who had been aboard the French freighter as she attempted the passage escaped in two lifeboats. Those in the first boat spoke nothing but French; strangely, those on shore at the time could not decipher their wild-eyed, frantic warnings. William

Mackie, a Halifax Harbor pilot and a bilingual Canadian, was in the trailing lifeboat; he spread the word in English when he got there. "Munitions aboard! Take cover. Run! Run!" he shouted excitedly to everyone within hearing.

Apparently, because of its military mission, the officers and men of the HMS *Highflyer*, a British cruiser, knew the truth about *Mont Blanc*'s cargo. Perhaps their vessel had been assigned escort duty to ensure that the needed munitions would arrive safely in Europe. It is otherwise difficult to explain the presence of a British military vessel in Halifax Harbor, so far from the battlegrounds that day. On observing the desertion of *Mont Blanc*, *Highflyer*'s captain immediately put a boat over the side. Three officers and twenty seamen instantly stepped forward to see if they could save the *Mont Blanc*. Their incredible bravery under the circumstances is difficult to comprehend. In the final analysis, their end was similar to that of every other sailor in the harbor who was above deck on that fateful day.

Mont Blanc's three-thousand-ton steel hull vanished in an enormous purple mushroom cloud. Sheets of flame and smoky, roiling turmoil shot up into the heavens at least a mile above. Slowly and ominously, the great cloud of gas enveloped outward toward shore, destroying everything in its path.

Sailors working their vessels' riggings were blown into the sea. Not only were they crushed, but contemporary accounts mention fierce fires burning away their clothes. One officer on a small vessel was blown half a mile away to a small hill, where he landed basically unhurt. His most severe problem was the fact that every shred of his clothing was blown away. On board the beached *Imo*, the unsuspecting captain and thirty of his crew were squashed like so many small bugs. Concern about sprung plates seemed minor compared to the devastation of the stacks, cabin, and towers. A dory operator crouched into the bottom of his small skiff. He and the boat were blown half a mile away to another section of the harbor. When the sailor regained his senses, he found he was unhurt, but thoroughly confused, floating along in his dory. Ironically, one of the French sailors who had fled was found dead fully a mile inland, pierced through by debris thrown from the ship. His coworkers apparently made it through to safety.

A half-ton iron fragment of *Mont Blanc*'s anchor wafted thousands of yards in on the rocky slopes. A huge, multiton rock wrenched from the harbor floor killed sixty-four people when it came to rest on the pier. Only those deeply below deck escaped the giant blow. The blast was so powerful it set up a thirty-foot tsunami wave within the harbor that wiped out most moored boats and all of the pier facilities in the region.

On the rocky shore surrounding the carnage, some five hundred small wooden houses were smashed into instant kindling. Many people were tossed as much as one thousand feet through the air. They came to rest against walls, rock outcroppings, and on top of buildings. Some miraculously survived, although many were stripped of most of their clothing and crushed by the concussion.

Automobiles, trucks, and even heavy trolley cars were thrown like toys into the hills. Thousands of fires started as a result of cast-iron cook and heating stoves being upended in the wooden houses. The fires added immeasurably to the confusion and desolation. Everywhere there was debris, as though some giant had come through and dusted the region with trash.

Giant leafless trees higher up on the hills were either snapped off or jerked—roots and all—out of the shallow rocky soil. Gulls and seabirds flying over the region were singed into ugly globs of flesh. Crushed, burned, and broken, they were instantly extinguished in Halifax Harbor.

Where the sugar refinery, Halifax elevator, and military gymnasium once stood, there was nothing. In some cases, not so much as a trace of these giant structures remained. Banks of warehouses—many full of goods—simply disappeared. Entire shifts of workers, many of them young girls, perished at their work stations in the many factories around Halifax. Heavy machinery was tossed about like small stones through walls, down hills, and into crowds of people.

Because so many people were drawn to the drama by *Imo*'s blaring horn, casualties were especially grim. Some estimates, which ultimately were alarmist, said that ten thousand people had been blinded. This occurred, it was rumored, as a result of them watching from inside through panes of glass at the instant of detonation. Eventually, the correct figure of two hundred blinded was released by the Canadian government.

People in the city of Halifax were somewhat protected by concrete and brick buildings and the region's hilly topography. Yet nearly every utility pole in the city was snapped like a toothpick. Even the underground water service was disrupted by the shock of the blast.

In Richmond, a small bedroom community to Halifax's south, all but three of the two hundred students in the school were killed when the brick walls blew in on them. (Unlike Dartmouth to the north, Richmond was never really rebuilt as a separate entity; eventually, it was absorbed by Halifax.)

The U.S. and British military presence in the region was fortuitous. American marines and sailors marched off the USS *Old Colony* almost instantaneously to assist with rescue operations. British naval personnel followed close on their heels. But many of the remaining homes, warehouses, piers, and factories were past saving. Most were allowed to burn through the night.

By midnight, however, a new threat developed. Halifax was in the path of the worst blizzard in recorded history. Temperatures plunged while snow piled up. Cruel, icy winds blew mercilessly over the maimed and the homeless. Many of those trapped in the wreckage were actually quick-frozen in place.

Overall, an estimated 500 people simply disappeared, apparently vaporized in the blast. Another 1,600 to 2,000 became fatalities. Property damage was estimated at more than fifty million dollars, simply an awesome amount for those times.

Very quickly, unbelievable amounts of aid started pouring into the region. In Boston Harbor, a relief ship was thronged with well-intentioned donors bearing food, blankets, and clothing. Special trains full of relief supplies were organized out of every New England state. Doctors with food, splints, chloroform, and bandages voluntarily hurried to the scene. The Canadian government, already strained from a three-year war effort, mounted a huge rescue effort of a scale unheard of prior to those times.

In the end, the total outpouring of support from all over Canada and the United States testified to the intensity of the event, a blast so powerful that it was even felt by some old, hard-of-hearing ladies, sitting at high tea on Prince Edward Island, 125 miles away.

Blast at German chemical factory devastates village of Oppau

OPPAU, GERMANY, September 1921—The great blast in Oppau, Germany, is among the world's most mysterious man-made explosions. It may also rank quite highly among the world's most stupid accidental detonations. The limited amount of available information easily leads to that conclusion.

The uncertainty regarding the incident isn't because the German borders were closed in 1921 when the little West German village of Oppau was wiped out in two successive blasts. The borders were open, in fact. The uncertainty seems to stem from fear and mistrust of a Germany fresh from a war treaty implemented in 1919, and from erroneous inferences made by the media.

Even sixty years after the blast, exaggerated references are still made to the 1,500 who died (there were actually 426 killed and 160 listed as missing), to the thousands hospitalized (there were, in fact, 1,952 injuries) and to the millions of tons of ammonium nitrate that detonated (an estimated 2,000 tons of explosives actually went up that day).

In journalists' eyes, Oppau was the benchmark ammonium-nitrate blast. It was a biggie, worthy of enrollment in the roster of the world's really great, free-ranging blasts, but whether it was the benchmark remains open for debate.

The suspicion surrounding Oppau was world class. Perhaps because of the time and the place, speculation as to what really happened ran very high. The Badische factory that was wiped out by the blast was the same one that first manufactured poison gas used by the German army in World War I. News commentators writing about the Badische complex disdained (somewhat tongue-in-cheek) "any certain knowledge that Germany was here again working out formulas and plans for high explosives and poison gases with a view to wars of the future." A reasonable translation might be, "We can't prove the Germans at Oppau were work-

OPPAU, GERMANY, Sept. 21—CITIZENS WANDER THROUGH WRECKAGE—In the photo on the previous page, townspeople walk through the smoldering streets after two violent explosions at the Badische Anilin und Soda Fabrik chemical plant. The blast destroyed the factory and most other structures in the small, densely populated company town, causing thousands of deaths and injuries. (Photo courtesy of The Bettmann Archive.)

ing on poison gas and military explosives, but we hope they were not thus engaged."

Commentators were universally ignorant as to exactly what was being undertaken at Oppau. One suggested, "Certainly the world would like to know more of what was being done at Oppau." Borders at the time were freely open. Any newsman who wanted to find out what had happened could have made an on-the-scene inspection. Instead, journalists speculated. "The company which owns the works is largely engaged in the manufacture of aniline dyes, and the manufacture of explosives and injurious gases is closely related to that industry," one reporter editorialized. No mention was made of fertilizers or agricultural chemicals, which are much more closely related to gases and explosives than fabric dyes. For those accustomed to dealing with America's advanced case of chemophobia, the entire episode had a familiar ring.

Although there apparently were no Western reporters on the scene, Western newsmen reported that, "The only possible way of rescuing the wounded was for firemen and other workers to wear gas masks, as in battle." The explosion, they said, originated in a reservoir containing two hundred tons of ammonium sulfate, subjected to tremendous pressures and very high temperatures, for the purpose of producing new poison gases for experimental purposes.

The fact that ammonium sulfate is not an explosive—even under the most severe circumstances—never seems to have entered these writers' heads. Virtually any college chemistry department in the United States could have set the record straight.

Accounts published in Germany several years later explained the incident in great detail. The truth, it seems, is grounded in such incredible stupidity that German authorities may simply have decided to allow destructive rumors to fly rather than own up to facts that would have destroyed the German reputation for being good, careful chemists.

Oppau was a very small, densely populated village right next to Mannheim in the Ludwigshaven district of lower Baden, in what is now West Germany. It was in this industrial complex that Badische Anilin und Soda Fabrik chemical works manufactured nitrogen products by compressing natural gas mixed with large quantities of air and water. Natural gas was produced from the soft brown coal found in abundance in the region.

The factory was originally built in 1911 to produce agricultural fertilizers, which were offered through normal channels to farmers, and also used as explosives. Homeowners in Oppau and the nearby city of Ludwigshaven also used surplus gas from the factory to heat and light their homes and to cook.

It was the designers' intent that the plant produce mostly ammonium sulfate as the fertilizer component. This material is very stable—virtually impossible to convert for use as an explosive. The explosive segment of the complex manufactured pure ammonium nitrate. Ammonium nitrate, in and of itself, is not generally explosive. It can, however, be the raw material used in dozens of very high-yield, military-grade explosives.

During World War I, Germany had experienced insurmountable difficulties importing the raw sulfur compounds necessary to produce sulfuric acid. Without sulfuric acid, it was impossible to combine ammonia and sulfuric acid to produce agricultural-grade fertilizer. Apparently, the resourceful Germans switched over to production of straight ammonium nitrate rather than bothering with the sulfate material.

All might have gone well if they had also switched from the mindset required for dealing with a relatively inert material to one necessary for dealing with one that is not so safe. However, additional complications soon surfaced.

As is true the world over, farmers only purchase fertilizer when they plant their crops. In the Northern Hemisphere, 90 percent or more of all fertilizer sales are made in spring and fall. Expensive manufacturing plants cannot be built to run only a month or two out of the year. As a matter of elementary economics, the factory ran year round, stockpiling production against the time of sales. A great number of sixty-foot concrete storage silos at Oppau were constructed to contain hundreds of tons of absolutely inert granular ammonium sulfate.

When the plant switched to the production of ammonium nitrate, frugal plant engineers in a somewhat poorly understood industry continued to store the product in the same sixty-foot silos that worked so well for the sulfates. Ammonium nitrate has several qualities that set it apart from ammonium sulfate, aside from its propensity under some conditions to explode violently.

For starters, ammonium nitrate is deliquescent. In a humid climate such as that of southern Germany, the chemical will quickly attract enough moisture from the atmosphere to become syrupy to pasty/sticky. Once it starts to pick up moisture, it will quickly and easily jump the next hurdle, setting up like green concrete.

At Oppau, given the great height of the silos, workmen were faced with an incredible handling problem. Column pressures compressed the material to the point that it would not flow out onto the conveyors built to unload the silos. Brave workmen even tried crawling into the silos through the bottom access doors in an attempt to dislodge the material with picks and shovels. Extreme danger from cave-ins made this a less-than-ideal procedure. The material commonly collapsed through empty, unconsolidated spaces, trapping and injuring workers.

In an almost desperate attempt to work out some way to sell their fertilizer, workmen hit on the procedure of blasting the consolidated ammonium nitrate down out of the silos by using low-grade, permissible dynamite. (Permissible, or safety dynamite, is even used in underground coal mines, where the danger of fire is very real.) Workmen drilled holes into the material with jackleg miners' drills, set the charges, tamped the hole, and shot the charge of dynamite inside the column of ammonium nitrate.

Records are somewhat vague, but apparently the procedure worked quite well until 7:32 A.M. on Wednesday, September 21, 1921, when the workers finally blew the whole town down. Two explosions occurred with such violence that the factory simply disappeared off the face of the earth. In its place, survivors found an eight-foot-deep lake about a third of a mile across. (While it reportedly has silted in a bit, the lake is still there today.) The first explosion was mostly a very large bang. The second sent a bolt of fire hundreds of meters into the sky. Immediately, a gray-black cloud spread like a heavy fog over the entire region.

Ruin was almost universal in the tightly packed village just rousing to the work day ahead. Oppau was comprised of about 6,500 people, most of whom lived within walking distance of their employers. Buildings in town were mostly three- or four-story vertical three- and four-plexes. Streets were narrow bands of cobblestone originally laid out for horse traffic. Many casualties occurred when buildings collapsed on unfortunate German workers crowding the streets on their way to work.

Some accounts speculate that casualties were minimized because only the workers blasting down the bridged chemical in the silos were present at the factory. In a small, tightly packed company town such as old-world Oppau, however, location was not a major consideration. Whether at home or work, everyone connected with the factory was in a place where the blast could reach them.

Under more modern circumstances, had residential setbacks been enforced, two thousand tons of relatively benign ammonium nitrate should not have posed a serious problem to bystanders. But in this case, the industrial town was so small and consolidated that when the factory went, it took the entire village with it. German roofs were fabricated from heavy clay tiles laid on support beams. Since they were not fastened down, loss of roofs in the region was virtually universal. Glass damage was reported as far away as Heidelberg, Worms, Darmstadt, and—in some cases—Frankfurt, forty miles distant.

Although there was little destruction of German towns during the war, the country was still of a mindset that allowed it to mobilize rather quickly to help. Ambulances were sent from as far away as Frankfurt. As Oppau had no medical facilities and/or doctors, most of the wounded went to hospitals in Mannheim. As a matter of record, French occupational forces were of great help supplying medicine of all sorts, as well as emergency food and shelter.

To some extent, the frangible buildings of Oppau cushioned the blast for the larger city of Mannheim. Mannheim had already evolved into the heavy manufacturing center it is today.

News commentators today, dealing from a position of basic ignorance, still refer to the lessons of Oppau in regard to the proper handling of ammonium nitrate. Perhaps among the piles of pony shit there really is a pony. It might be that one should not pile ammonium nitrate sixty or more feet high, creating severe pressures below. And perhaps one should not attempt to shake the material down onto a conveyor using dynamite.

The greatest lesson probably stems from the fact that nothing nefarious was being undertaken at Oppau. Just plain, old-fashioned, grassroots stupidity.

Series of blasts set off by lightning wipe out U.S. military arsenal

LAKE DENMARK, NEW JERSEY, July 1926—Lake Denmark was, for the United States, much the equivalent of the Severomorsk blast for the U.S.S.R. Like the Russians, the United States lost virtually all of its fleet's munitions to the capricious acts of a free-ranging explosion. Damage was extensive enough to destroy roughly a third of the nation's yearly budgeted allotment of military munitions. Local commentators pointed out that it was extremely fortunate that the country was not at war at the time. "Should the military have immediately required the munitions, the results of Lake Denmark would have been catastrophic," one editorialized.

The blast was the last really grand, pure military explosion experienced in the United States. Perhaps it only took one such blast to teach the necessary lessons regarding the storage and handling of high explosives.

Lake Denmark Naval Ammunition Depot was near

the small village of Dover, New Jersey. Dover is located in the top fifth of the state, about thirty-five miles due west of Paterson. Numerous lakes dot the rolling countryside, which is covered by a dense growth of scrub oak and elm.

U.S. naval vessels would sail up the Hudson River past New York City and berth in the Fort Lee vicinity to take on munitions. In that era, the vessels were loaded principally by rail. The army's smokeless powder factory, Picatinny Arsenal, adjoined Lake Denmark in a remarkable example of interservice cooperation. Some thirty buildings, including several large manufacturing centers, comprised this complex, which for all practical purposes blended into one large base. Lake Denmark (the actual lake) lay north of the arsenal and depot, snuggled into the gently rolling north Jersey sandhills. In 1926 dollars, the entire complex was said to be worth about $87 million.

Captain Otto G. Dowling, commander of the sprawling Lake Denmark Arsenal, sat on the second-story sun porch of his base home with his wife watching a spectacular midsummer thunderstorm roll northeastward across the picturesque green countryside. At 17:15, Dowling remarked to his wife that an unusually wicked-looking bolt had struck at or near temporary magazine No. 8. Captain Dowling knew that No. 8 contained well over a million pounds of high-intensity TNT cast into charges appropriate for depth charges and aerial bombs. Even though the building was temporary, it was protected with a state-of-the-art lightning-arresting system. Yet, shortly after the strike, the couple spied a thick column of greasy black smoke pouring out of the south end of the building. Almost instantaneously, their phone rang with a report from the on-duty officer. "Fire at or near the west gate," he screamed in panic. Without hesitation, Captain Dowling grabbed a raincoat and ran to his car.

Colonel N.F. Ramsay, the ranking army officer at the adjoining Picatinny Arsenal, heard the fire alarm from his office little more than half a mile away. Desperately, he ordered the telephone operator to try to connect him with someone at Lake Denmark. His efforts were singularly unsuccessful. He walked back over to the window just in time to look out and see an enormous white flash. He instinctively dropped to the floor. A few seconds later, the concussion reached his office building, blowing in the side like it was made of toothpicks. Ramsay now knew that the situation was out of control and that use of the phones was out of the question.

Meanwhile, Mrs. Dowling ran down the stairs after her husband. As she passed through the kitchen, the first blast crinkled the house. She was thrown to the floor amongst shards of window glass and fractured plaster. Getting up, she ran out the front door. A second blast hit and completely crumpled the house just as she reached the driveway. Behind her, the roof dropped in, no longer supported by functional walls.

A fence surrounding their compound fell flat as the walls of Jericho. On the street, Mrs. Dowling turned to see a wall of flame rolling down the hill toward her, emanating from the exploding magazines. With excellent presence of mind, perhaps generated by her long association with the military, she turned and ran toward a wooded area behind the house. The recent rain had thoroughly soaked the foliage, and she reasoned that it might provide some cover from the fire storm. Before reaching the small wooded area, she was thumped several more times by exploding ordnance (apparently, the high-explosive warheads were being thrown into other nearby storage buildings, whose contents were, in turn, detonating). At the woods, Mrs. Dowling found she was accompanied by a number of additional refugees similarly fleeing the holocaust. She ran eight miles through oak and sumac thickets, tearing her dress to shreds. At the village of Green Pond, sympathetic neighbors gave her a replacement garment to cover her cut and torn body.

Hapless visitors driving through the depot at the time of the first blast found the roads covered with debris and, in some cases, live rounds of up to six-inch ammunition. Several autos were whooshed accordion-style into unusable heaps of tin.

The entire area was puffed clean of all buildings, people, vehicles, and telephone lines. Spectators at the base, mostly veterans of the war in France, found the rolling detonations to be frightfully similar to their personal experience in Europe. Many jumped uncontrollably whenever yet another high-explosive round detonated. (Because black powder was used extensively at the time as a powder booster in the navy's large nine-, twelve-, and fifteen-inch guns, a number of the multiple thumps originated from this source.)

Fire fighters at military arsenals seem to possess a fatalistic outlook toward life. Five post fire fighters responded immediately at the first alarm. Fearlessly, they raced their doomed truck toward magazine No. 8. (Had they been a bit less efficient, they might have lived.) Ten hours later, nothing, not even the motor block, stood more than fourteen inches high. The tires were even separated from the steel rims.

Blasts and burning continued unabated for almost ten hours. Each magazine in line detonated, throwing burning shells into the next building. Most were smashed to pieces by the massive concussions, becoming instant kindling for the burning ordnance. Of the post's 180 buildings, only 16 survived. Thousands of tons of expensive explosives went up in smoke. Every building within 2,700 feet of the arsenal was swept off the face of the earth. Even beyond that, buildings as far out as 8,000 yards were severely realigned.

Hospital building No. 49, for instance, lay in ragged ruins. Medicine bottles and broken water pipes were scattered over the tilted floors, mixed with bent, smashed X-ray machines and radio sets. It was a sorry sight for those who appreciated the fact that state-of-the-art medical equipment had been trashed before any productive use could be made of it.

Although an occupied marine barracks lay in the path of the blast and a number of visitors were passing through the base that Saturday afternoon, casualties were surprisingly light. Final figures suggest that not more than thirty people were killed. About two hundred wounded were treated at hospitals in Morristown and Dover, fifteen to twenty miles away.

One marine doing KP duty that afternoon was thrown fifteen feet across the kitchen. He ran some three miles before realizing the initial blast had blown away his boots. Bloody and crippled, he wrapped a shirt around his feet so that he could continue his retreat.

Residents as far as fifty miles away reported hearing or feeling the blasts. Ten smaller villages, including Rockaway, Marcella, Green Pond, and Mount Hope, were damaged. Some were impacted so severely that they had to be rebuilt from the ground up. Citizens of Mount Hope, an early iron-mining village accustomed to the fickle nuances of history that brought about alternating periods of prosperity and poverty, typified the reaction of other citizens within reach of the blast. Most simply abandoned their plain little houses, running or driving away into the countryside.

Rockaway was the center for the regional telephone exchange. Although the village suffered extensive damage, the "hello girls," as telephone operators were then called, elected to remain at their stations and run what remained of the local telephone system. Their bravery under extreme duress gathered accolades from the many citizens who jammed the circuits with anxious calls. Reportedly, they calmly handled ten times the normal volume of calls from separated family members and others wondering what was going on. In Dover, ten operators stuck to their stations for twenty-four hours straight.

By Sunday morning, the navy had assembled 220 combat-ready marines from Quantico, Virginia, to fight the remaining blazes. Working with World War I gas masks and tin derbies, they were often able to save quantities of valuable large-bore ammo. Some marines reported that every time another fifteen-inch shell detonated, they could feel it up into their knees through the soles of their boots.

When it was all over, in addition to the complete loss of the $87 million arsenal and powder factory, surrounding civilian property incurred an estimated $70 million in damage. Personnel injuries were minimized, since workers labored only half a day on Saturday. Damage to the Picatinny Arsenal buildings was extensive. None of the smokeless powder or ingredients burned or exploded, however, largely due to the generous spacing of buildings and their somewhat protected location. Five-inch shells from the naval depot had been thrown over the hill onto the army's nine-hole golf course, rendering it completely unusable. It looked as though a giant duffer had walked through, slicing out multiple two-foot divots. The War Department classified the Lake Denmark blast as probably the worst in the history of the world and certainly the worst in the history of the United States. Public reaction to the event suggests that the ideas of today's antinuclear activists are far from new or original.

Residents of the region complained about being subjected to the rigors of war in peacetime. Great, deep craters were torn in the loose, sandy New Jersey hills.

Many hills were now piebald, much like places citizens had seen in pictures of battlefields. Steel girders were thrown almost a mile from the blast site. In one case, a chunk of charred wooden beam turned up on a farm three-and-a-half miles away.

The *New Haven Register* editorialized, "To say the tragic events of Saturday night were due to an act of God in sending a bolt of lightning upon the planet is all right as far as legal liability may be concerned. Acts of God and the common enemy are ever good defenses in courts, but before the bar of public opinion, the persons responsible for the storage of wartime supplies of deadly explosives in peaceful industrial and residential communities or near enough such communities to kill and destroy are culpable."

New Jersey's governor and its two senators called for the army and navy arsenals to be removed from the state immediately. Although most newspapers supported the fact that the incident was truly an unavoidable accident, they also supported the governor. The media generated rhetoric that sounds vaguely familiar today; one paper suggested that "New Jersey had lost more people from powder explosions and the blowing up of ammunition plants than have been killed in many of the important battles of the Revolutionary War and in the Spanish-American War."

While the statement is misleading—and, at best, only partially true—in the case of Lake Denmark the fact remains that, due to its volume and duration, the incident was surely one of the great free-roaming blasts of history. New rules of procedure were formulated, and arsenals were spread out and located to more remote regions of the country where the population had not reached dense levels.

The United States, like the Soviets sixty years later, learned a very expensive but important lesson regarding the storage and handling of ordnance.

School blast kills 455 students, teachers in New London

NEW LONDON, TEXAS, March 1937—A few very common materials with which we live very closely can be extremely explosive. Specific or—depending on one's point of view—random circumstances can allow these materials to detonate quite easily.

American GIs, for instance, were often treated to a demonstration wherein a demolition team using fogging devices treated a deserted building with atomized gasoline. When this mixed with oxygen and was detonated with a standard No. 6 blasting cap, the building was whapped right off the map.

A gallon of gasoline, the old saw goes, has an explosive equivalent of about seventy-five sticks of dynamite. Like gasoline, natural gas can also be fairly explosive under some circumstances. One Texas afternoon in 1937, 690 students learned this lesson the hard way. At 3:05 P.M. on Tuesday, March 18, almost a third of them succumbed to what probably still stands as the world's greatest schoolhouse blast.

(Many a student has conspired to blow the school up while in chemistry lab.) But this explosion—while triggered by a student—was much more complex in origin

NEW LONDON, TX, Mar. 20—RUINED SCHOOL BUILDING—Photo on previous page shows one section of the wrecked Consolidated School following the natural gas explosion that killed more than 450. The photo above was taken from a similar perspective prior to the explosion. (Photos courtesy of Archives Division, Texas State Library.)

than simply blending charcoal, sulfur, and potassium nitrate together in the lab.

A student named Barber, who ultimately scrambled to safety through the trashed structure, claims he heard a sound "like dynamite" that caused the roof to lift up slightly, allowing the walls to fall in, and then dropping the heavy roof back down on top of the whole mess.

An unnamed oil-field tool pusher, along with his crew of roughnecks working nearby, was probably more able to identify the sound. "Sounded like a firecracker in a can," he told a reporter. "It was a muffled pop that simply blew the sides out and the roof up."

In reality, it was rather muffled, a gentler and kinder explosion, the intensity of which is grossly overstated by the resulting 455 casualties. Judging by the condition of most of the nearby buildings, it could not be classed as a devastating blast, certainly not in the league of Halifax, twenty years prior, or of Texas City, yet ten years in the future.

The roofs and windows of most nearby buildings remained basically intact. One notable frame structure within two hundred yards survived the blast with all of its many glass windows on the exposed side still whole. Yet the blast had moments of authority, as when it threw a giant ton-sized slab of concrete across the street onto a teacher's parked automobile.

Photos taken minutes after the blast show rows of huge luxury-class Chryslers, Packards, Cords, and Cadillacs, many—including the open touring cars—dusted with debris. Oil derricks covered the surrounding countryside, blending well with the pricey automobiles parked around the blast site.

Moments after the blast, the grassy play area next to the school looked as if someone had tossed a basket of rag dolls about. Smashed and broken bodies of what seconds earlier had been normal high-school students littered the ground. Officials and parents pitched in

NEW LONDON, TX, Mar. 18—BLAST TAKES HUNDREDS OF CHILDREN—The above photo shows the wreckage of the New London Consolidated School, where hundreds of students were crushed to death in a violent explosion. Ambulances were rushed in from all surrounding cities. Shown below is a comparable view of the school prior to the blast. (Photos courtesy of Archives Division, Texas State Library.)

with great bravery, doing whatever they could. Ironically, some one hundred parents were gathered across the schoolyard at a PTA meeting when the blast occurred. Terminal boredom turned to intense critical involvement in a matter of seconds.

At 3:05 P.M., when the detonation occurred, only some 690 students plus 40 teachers were in the ill-fated building. The grade-school students had been dismissed earlier. Had fate delayed the blast another ten minutes, the remainder of the students would have been merrily on their way home as well.

Paula Echols, a seventeen-year-old student who was sitting in English class, survived the blast. She recalls hearing a kind of whomp, feeling the building shake, and then, the next thing she knew, seeing a teacher's leg protruding out from the bottom of a pile of brick and mortar.

At the time of the blast, the New London school district was reportedly the wealthiest in the United States. Royalties from local oil and gas wells contributed more than a million dollars a year to the school's maintenance and operations. Programs and facilities were considered to be among the most advanced in the nation. References were often made to this being a "model school system."

From the first, there was little doubt as to the cause of the explosion. Officials maintained that explosive gas had seeped up naturally from the ground below the building. At least six oil derricks covered the school property, some as close as three hundred yards to the pile of rubble that once had been a school, lending credence to the fictitious explanation.

Perhaps because of the highly emotional nature of the casualties, the investigation of the incident took some very strange twists. William C. Shaw, the school's superintendent, was quizzed for three days in a manner that sounds much like something carried out by security police in some Third World country. After three days of intense questioning, Shaw finally admitted that he had surreptitiously tapped into the Parade Gas Company's wet gas line. The line carried waste gas from nearby oil wells to a place where it was flared harmlessly away.

Wet gas is considered to be highly explosive under virtually any circumstances. The presence of water vapor dissolved under high pressure in the gas adds oxygen and hydrogen in a manner that is very volatile. Additionally, natural gas straight from its underground source has no odor. Until it is purified by dewatering and an odorizing agent is added, the material is most difficult to contend with.

Superintendent Shaw eventually admitted that he had tapped into the gas line as a cost-saving measure to eliminate an estimated $250 annual heating bill for the school district. He maintained that Parade Gas was aware of what he was doing. (Parade steadfastly maintained that no approval had been solicited or extended. No written documents were produced as evidence for either side.)

The school board was also implicated for its penny-pinching posture in refusing to install a central heating system. It had elected to use individually fired room heaters—a most curious decision for the best-endowed school system in the nation. (New London, Texas, lies about one hundred miles east and a bit south of Dallas—not an area where trustees normally expend a great deal of money on heating facilities. By mid-March, one would expect cooling to be of greater consideration than providing additional warmth.)

Subsequent investigation by Dr. Price, an expert from the Texas Bureau of Chemistry and Soils, indicated that a significant gas break was created when maintenance personnel repaired one of the individual in-wall gas heaters. Superexplosive wet gas was

NEW LONDON, TX, Mar. 19—TWISTED WRECKAGE SCANNED FOR BODIES—Following the disaster, workers scan the twisted masses of concrete and steel for bodies of children killed in the Consolidated School blast. (Photo courtesy of Archives Division, Texas State Library.)

NEW LONDON, TX, Mar. 19—TEXAS SCHOOL BLAST—Crowds converge upon the campus to view the damage to the rural school that distinguished itself as "the richest school in the world." (Photo courtesy of Archives Division, Texas State Library.)

allowed to drain out of the faulty fitting into the inner wall, where it flowed down into a crawl space beneath the one-and-a-half-story rural schoolhouse.

John Nelson, a seventeen-year-old shop student, apparently blew his school down when he flipped the switch on a bench sander. Nelson, who survived the incident, said he heard a huge whomp, and a mass of sand and soil was propelled toward him by a giant fireball.

Most of the casualties were caused by falling brick, beams, and plaster. Few, if any, of the 455 students who died were actually rent in pieces by the blast. At the time of the incident, the New London community was just completing a brand new hospital. Dedication ceremonies were contemplated, but no official opening had taken place. One can but wonder at what levels the facilities were staffed, but the hospital opened ahead of schedule to receive the flood of victims. Doctors ran out of ether, chloroform, bandages, and splints. Downtown, the community ran out of coffins. Many victims were interred in rough wooden boxes built by volunteers at the site.

Repercussions from this emotional incident bounced around the state for years. New laws and inspection procedures were instituted, and standards were toughened all around. Because the blast did not destroy a vast section of the surrounding community, and perhaps because of the public's attention to the beginning of U.S. intervention in World War II, the nation now lacks a vivid recollection of the incident.

Fire storm rolls through east Cleveland neighborhood following series of gas explosions

CLEVELAND, OHIO, October 1944—The United States was in the final throes of World War II when the East Ohio Gas Company detonated on Saturday, October 21, 1944. The war had not yet been decisively and absolutely concluded, but Nazi Germany was in retreat, pressed on either side by cooperating Allies. Elements of the Soviet army had routed the best German forces in the Battle of Kursk, the largest set-piece battle ever fought. Time was running out for the Japanese as well, but neither they nor the United States could forecast the speed with which the conclusion would arrive.

From a late 1944 vantage point, it looked as though the war would continue at least several additional years. Massive numbers of Americans continued to be summarily torn from their chosen professions in their home communities and sent to places most had never heard of. Women and older men judged unfit for service were left behind to operate the machines in factories, milk the cows, and run the fire departments.

Rationing was still in effect. Shortages of materials we now take for granted were endemic. Even prestigious national news magazines and papers were allocated only a limited amount of supplies with which to complete their work. Newspapers, faced with remarkable news stories, did not have the option of simply printing additional pages. East Ohio Gas Company's catastrophe was remarkable in the immediate havoc it created and its long-term effects on fire codes relative to natural-gas storage and transportation. It was, however, mostly ignored by the media, constrained by a lack of paper and ink.

One account included a single photo of the incident, along with a three-line caption. The caption dwelt on the fact that the explosive fire storm destroyed hundreds of cars parked in a nearby lot. The level of concern for the hundreds killed and injured seems remarkably low in retrospect. But automobiles were a scarce commodity in the United States in 1944. For the most part, they had

CLEVELAND, OH, Oct. 24—BIRD'S-EYE VIEW OF BLAST AREA —On the previous page is an aerial view of the area gutted by an explosion and fire on Cleveland's industrial east side. More than 100 persons were sent to hospitals and 135 were reported dead from the blast. (Photo courtesy of The Western Reserve Historical Society, Cleveland, OH.)

not been manufactured since 1941.

East Ohio Gas Company first built their natural-gas compression and storage facility on the east side of Cleveland in 1940. Liquefied natural gas was still something of a novelty in 1944. Few people seemed to realize what was involved with the manufacturing process or the dangers involved. Local authorities apparently raised few if any objections when the city moved in eastward around the tanks. It wasn't a safe situation, but policies and regulations regarding such matters were poorly defined in the United States at that time. When first filled after construction, one of the three 240 million-cubic-foot tanks leaked slightly. It was repaired and put into service with little thought regarding possible long-term consequences.

Shortly after noon on Saturday, October 21, one of the tanks sprang a sizeable leak. It began by squirting supercold gas under intense pressure out into Cleveland's humid atmosphere. Observers said the leak grew, creating a thick fog that soon blanketed the area. Once mixed with oxygen, such gas becomes extremely explosive.

Emergency repair workers were summoned to restore the tank. Transferring the gas to an alternative tank was impossible; all were filled to capacity in anticipation of Cleveland's approaching heating season. The only way a tank could be emptied was by being drawn down by industrial and residential users.

Bravely—or perhaps stupidly—the company repairmen gathered to repair the leak. Suddenly, the gas reached a spark or open flame of some sort. Contemporary accounts did not specify the exact cause. No doubt there were many possibilities. It could be that in a pressing wartime situation, the authorities that were available to investigate did not discover the exact cause of the explosion. All who were involved may have perished. Censors may, for some strange rea-

CLEVELAND, OH, Oct. 20—FIRE LIGHTS THE SKY—Onlookers stand in awe as flames from a burning storage tank illuminate the sky. Spreading fire drove more than 1,500 people from their homes following a multimillion-dollar explosion at the East Ohio Gas Company. (Photo courtesy of The Western Reserve Historical Society, Cleveland, OH.)

son now lost in history, have arbitrarily dropped a curtain of secrecy on the event. Perhaps they felt that releasing more than a very few basic details would compromise the war effort.

Like most gas explosions, this one was more of a thud than a blast. Mrs. Charles Flickinger, in her home a block from the tanks, walked over to the wall to plug in her vacuum cleaner. (Saturday was her day home from the factory, when she tried to catch up with her household chores.) At that instant, she heard a muted, muffled explosion. The walls of her home turned a cherry, charred red, and her curtains broke instantly into flame. She ran immediately out into the street, joining the crowd of people hurrying to evacuate the area. She thought only of escape, losing everything but her life in the incident.

Other, more severe blasts followed as the remaining tanks heated and detonated with considerable force. As the explosions spread through the region, they took out all of the large buried gas mains, blasting block after block of the city into unusable rubble. A survivor out on the streets looked out across a factory roof to a chimney and calculated that the flames briefly reached an altitude of 2,800 feet. Pigeons, roasted in midair by the intense flash of heat, fell to the pavement, adding to the grotesqueness of the situation. Heavy cast-iron manhole covers shot up and out onto the parkways for blocks in every direction like massive flying poker chips. There are no records of injuries directly attributed to these flying chunks of cast iron, but some must have resulted among the people running down the streets.

The massive fire storm rolled through the neighborhood, destroying more than 165 homes. To some extent, many residents were able to flee the fires. But the tragedy left more than 1,500 homeless in a very house-poor America. Block after block of houses were

CLEVELAND, OH, Oct. 20—ACRID SMOKE SPREADS—Smoke blackens the sky on Cleveland's east side as fire spreads following the explosion of two liquid gas storage tanks. (Photo courtesy of The Western Reserve Historical Society, Cleveland, OH.)

CLEVELAND, OH, Oct. 24—DEVASTATION SCENE—Vast area of devastation in east Cleveland, where a huge explosion caused an umbrella of flame to cover several blocks, igniting dwellings and other buildings. (Photo courtesy of The Western Reserve Historical Society, Cleveland, OH.)

CLEVELAND, OH, Oct. 20—WALLS COME TUMBLING DOWN —Police officers look on as building crumbles in the wake of the industrial explosion. (Photo courtesy of The Western Reserve Historical Society, Cleveland, OH.)

scorched and smoked and went up in flames. At day's end, the site looked much like it had been firebombed. Only brick and block chimneys still stood amongst the blackened, smoking ruins.

The old men who staffed the fire department did their very best. They quickly set up a perimeter from which they sought to contain the fire. They poured millions of gallons of water onto the blaze. More sophisticated techniques were not available at that time.

Then, as though they were fighting a war, the fire fighters started advancing on the enemy gas tanks, long ruptured and bent into grotesque shapes by the heat. As they advanced, they encountered body after body, burnt beyond recognition. The press, perhaps hardened by the brutal war, referred to the 135 casualties as people who "crisped and died like moths."

CLEVELAND, OH, Oct. 23—WORKERS CLEAR DEBRIS—Two men take on the task of clearing the rubble. This is all that remains of a house near the East Ohio Gas Company. (Photo courtesy of The Western Reserve Historical Society, Cleveland, OH.)

CLEVELAND, OH, Oct. 21—DEAD RECOVERED FROM RUINS—Emergency rescue personnel were too late to save this victim—one of the 135 reported dead in the blast. (Photo courtesy of The Western Reserve Historical Society, Cleveland, OH.)

People in east Cleveland who lived adjacent to the East Ohio Gas Company storage yard lost everything. Their homes might just as well have been located in France or Italy. The devastation was equally thorough.

Bulldozers were brought into the area to level buildings or parts of buildings still standing. New safety rules were written prescribing the testing of tanks, use of safety valves on mains, and residential setbacks (or clear-zones). For instance, no building was allowed to remain adjacent to a gas storage site. These and other measures quickly became standard across the United States.

As perceived by officials away from the immediate zone of destruction, the biggest single problem related to the blast was the impact on war production. National papers reported with absolute horror that a total of thirty-nine vital Cleveland war plants had to shut down due to a lack of fuel and/or means of delivering it. It was an event of great concern for a nation still struggling mightily to support their young men at the front.

Ship explodes at Monsanto pier; turns Texas City into virtual battle zone

TEXAS CITY, TEXAS, April 1947—The huge, weathered public clock on Michaels Jewelry store stopped stone-cold dead at 9:19 A.M. on Wednesday, April 16, 1947. For years, it had stood as a symbol of reliability and honesty for the store owners below, as well as a customer attraction. Neither the Texas heat and humidity nor the violent, unpredictable hurricanes of the thirties had been able to deter the mechanism from its appointed rounds. It stood on vibrant Main Street, a symbol of progress in this burgeoning community.

Louis Alexander, the faithful and unsuspecting store manager, arrived at work a bit early on that fateful morning. His ambition was born as much by the prospect of avoiding the wet, sweltering heat that was building out over the bay as by the desire to cash in on the Texas City economic boom. (During the war years, farmers and ranchers had flocked to the relatively easy, high-paying jobs the city offered.)

Alexander and other downtown merchants, fresh from their look-alike homes built on perfect suburban squares, were unaware of the drama unfolding three miles south of town at Monsanto pier.

But 3.2 million pounds of ammonium nitrate has a way of catching people's attention when detonated all in one pile. Those who did see the blast and survived claimed the mushroom cloud was Bikini-like. Not quite two years had elapsed since the United States had used its first two atomic bombs to end a world war. Nuclear tests continued, many of which were openly publicized. Stereotyped pictures of mushroom clouds over the test range on Bikini were vivid in people's minds. A mushroom cloud rising four thousand or more feet across the bay certainly would have reminded people of the blasts they saw pictured in the papers. (The cloud may have projected even higher into the heavens, but the dense, overcast canopy at four thousand feet precluded mere mortals from knowing.)

TEXAS CITY, TX, April 18—BLACK CLOUD OVER TEXAS CITY—In the photo on page 87, smoke blackens the sky over the burning Monsanto Chemical plant in the disaster that took more than five hundred lives in Texas City. The explosion of the nitrate-laden ship touched off the tragedy, which destroyed two-thirds of the city and caused damage estimated at $125 million. (Photo courtesy of *Texas City Sun*, Texas City, TX.)

Alexander walked across the main lobby of his newly refurbished store on orderly, buttoned-down Sixth Street in downtown Texas City. Although it was technically still only springtime, some of the northern visitors were already commenting on the oppressive heat. (Texas City sits at twenty-nine degrees north latitude, on the same line as Orlando, Florida, and Chihuahua and Hermosillo, Mexico.) He thought about the advisability of opening earlier and closing later to accommodate his customers and avoid the coming swelter. He did not have time to dwell on the issue. As he turned toward his desk, the heavy front door snapped off its hinges like a fractured saltine cracker. Shards of glass showered into the room; Louis Alexander was pitched headlong across the store like a rag doll.

Fertilizer-grade ammonium nitrate, thought by most experts to be mostly benign, was not viewed with much alarm in those days. Quarry operators used it as an inexpensive substitute for dynamite by soaking it in fuel oil, and even then the mixture often required one third or more dynamite to detonate. (Knowledgeable kibitzers know that—even at worst—ammonium nitrate is a stable, hard-to-detonate material. When used as an explosive, it replaces relatively mild 40-percent commercial dynamite.)

Thus, in the case of the 110-foot French freighter *Grandcamp* tied to the Monsanto pier, no one was more than routinely alarmed. Even when normally ominous warning signs started to materialize, most Texas City residents—like Louis Alexander—continued on with their normal routines. They did not read the signals as portending anything particularly threatening.

At 8:00 A.M., smoke was seen coming from the freighter's number-four hold. To this day, no one is sure of the origin of the fire. A Coast Guard inquiry concluded that an errant cigarette set it. (Looking back, this assessment seems more of an indictment of "the

TEXAS CITY, TX, April 18—WRECKAGE IN DOCK AREA—A view of the dock area in Texas City after the 110-foot French freighter *Grandcamp* exploded at city docks, causing widespread damage from resulting fires and detonations. The ship was loaded with 3.2 million pounds of ammonium nitrate. (Photo courtesy of *Texas City Sun*, Texas City, TX.)

TEXAS CITY, TX, April 18—CHEMICAL PLANT IN RUINS—The gutted Monsanto Chemical facility following the violent explosions that rocked Texas City. In the foreground is a ship's propeller, which was hurled several hundred yards inland by the force of the blast. (Photo courtesy of *Texas City Sun*, Texas City, TX.)

dreaded weed" than any real explanation for what happened.) An engineer secured the cargo hatch and opened a steam valve, flooding the compartment with live steam in an attempt to control the fire. Had the cargo been dimension lumber or paper, this measure might have been effective and wise. As it was, the heat and moisture set fate's detonator timer running.

About 8:30 A.M., the Monsanto in-house fire team concluded they were fighting a losing and perhaps perilous battle. Orders were passed to abandon the ship, tow her out eleven miles toward the Gulf, and clear the area of the many hundreds of spectators who congregated to watch the show. Those orders were only partially executed.

The resulting blast was severe enough to register on the seismograph at Denver, Colorado, nine hundred miles north. Observers later estimated that four airplanes were blown from the sky by the concussion. It was the era of Stinson Reliants, Piper J-3s, Cessna 140s, and Luscoumbes. Few had radios. During the ensuing melee, many others may not have returned to base, knocked from the sky by the concussion. No one knows for sure.

A solid-steel propeller drive shaft some sixteen inches in diameter, perhaps twelve feet long, weighing thirty or forty tons, was hurled hundreds of yards inland. It came to rest in a railroad-marshalling area, stuck in the ground at a grotesque angle. It was as though some giant was throwing huge spike nails about.

Low-flying ducks, seagulls, herons, and pelicans were knocked from the sky by the hundreds. In many places, they cluttered the ground like scraps of waste paper blown from a garbage truck. Damp, oppressive temperatures quickly deteriorated the carcasses, adding to the pall that had begun to hang over the region. Later detonations would come, but the birds were past caring.

Virtually everything man-made within eight to ten miles was trashed instantly. Broken boards, bricks, and shards of steel covered the ground everyplace, impeding the movement of emergency vehicles. What had been a relatively new, orderly urban area became a battle zone.

Although not yet fully destroyed, the Monsanto Chemical plant, with its many storage tanks, was in the process of self-destructing. Built at a cost of nearly $125 million to support the war effort, the tank farm and dock covered what today would be considered a modest thirty acres. Ammonium nitrate with export potential was produced in quantity at the plant, using abundant natural gas from Texas oil fields as feed stock.

Fires burned on for all of that day. Tanks containing ammonia, propane, styrene, and benzol broke into flames, adding to the incredible heat. Greasy black smoke blanketed the region. Remaining city and company firemen who stood bravely at their posts knew without a doubt that they were not playing on the winning team.

As is usual in these circumstances, a few citizens reacted with great bravery, performing superhuman tasks. Frank Taylor, a Monsanto employee, was blown through a hole in a thick cement wall of his office at a warehouse. He fell six feet into the bay and swam half a mile across it to the place where his home once stood. Nothing remained that he could recognize. Others wandered aimlessly through the ruins, mumbling incoherently. Some purposely walked into the inferno and died. A Mrs. Lida was tossed through a window from the second floor of her office by the initial blast. The same concussion that blew her out the side of the building served to cushion her fall. She survived this and the resulting fires that ravaged the

TEXAS CITY, TX, April 18—GRIM AFTERMATH OF TEXAS DISASTER—One of the many structures twisted and flattened beyond recognition by the explosion that turned Texas City into a virtual battle zone. (Photo courtesy of *Texas City Sun*, Texas City, TX.)

TEXAS CITY, TX, April 18—FIRES RAVAGE CHEMICAL PLANT—What wasn't destroyed by the force of the initial blast was eventually ravaged by the resulting inferno that spread throughout Texas City. Shown here are the skeletal remains of the Monsanto Chemical plant. (Photo courtesy of *Texas City Sun*, Texas City, TX.)

region over the following days. Hundreds of others were hurt by glass blown to dust-fine shards.

That night, the conflagration spread to the Republic Oil refinery and to South Port Petroleum and Humble Oil. It was impossible for even seasoned employees to know where familiar buildings, tanks, and parking lots once stood.

Fires and explosions continued through the night and into the next day. Monsanto officials correctly maintained that their tanks full of chemicals were not explosive; however, they were highly flammable. Relief workers, including the famous Texas Rangers and the Texas National Guard, began arriving on the scene.

Two additional ammonium-nitrate carriers, the *High Flyer* and the *Wilson B. Keene*, looked imperiled. The *High Flyer* was loaded with fertilizer. The exact contents of the *Wilson B. Keene* have been obscured by time and the events of the previous day. At about noon, orders were given to tow both vessels out into the bay. Wary tugboat operators managed to think of urgent business elsewhere. Brave rescue workers did the next best thing by cutting the vessels loose from their mooring, hoping the tide would carry them out. The measure provided only a small amount of benefit.

When the *High Flyer* disintegrated at 1:11 P.M., it took the *Wilson B. Keene* with it but contributed little more to the general mayhem. (It was like swallowing a live toad. Once that is done, nothing else really bad can happen to you that day.)

One Major J.C. Trahan, wounded in a Nazi blitz in Europe, later said he had never previously experienced anything like what he saw going on around him. Residents in Galveston, ten miles across the bay, reported eardrums and plaster walls shattered with similar impunity. Whole rows of houses were washed into compressed, pancake-like vertical stacks. Money

TEXAS CITY, TX, April 18—SUNKEN SHIP—The *Wilson B. Keene* was one of two additional ammonium-nitrate carriers at Monsanto pier that was destroyed when fires resulting from the explosion of the *Grandcamp* touched them off. (Photo courtesy of *Texas City Sun*, Texas City, TX.)

TEXAS CITY, TX, April 18—GUTTED TEXAS CITY HOME—Some homes in Texas City were flattened like pancakes by the force of the blast. This one remained upright, but the walls caved in. (Photo courtesy of *Texas City Sun*, Texas City, TX.)

deposited at the First State Bank downtown was strewn about the street. Diligent clerks tried to police it up, but generally, the bounty was ignored by citizens, who faced more pressing tasks.

With communications devastated, many citizens were still uncertain of the magnitude or even the cause of the travail under which they labored. Rumors spread among people anxious for some news—even if adverse. There were rumors of additional vessels loaded with ammonium nitrate, military munitions, ruptured gas mains, and other such doomsday prophecies. Rescue workers donned World War I-vintage gas masks as a precaution against poisonous gas reportedly created by the burning chemical tanks. There were some threatening chlorine-gas and nitrogen-dioxide clouds, but mercifully, these blew out to sea on a southeast wind.

Other than direction provided by the Texas Rangers and the National Guard, governmental function had been destroyed. The few fire fighters that remained were cut down like wheat when the *High Flyer* blew.

By afternoon of the second day, things were quieting down slightly. Most of the fires had consumed the bulk of available fuel. Without new sources of combustibles, die down was inevitable. Nonetheless, jittery, shell-shocked residents spread rumors about additional impending blasts. Confusion and uncertainty ran rampant.

Even in the sweltering climate, clothing and shelter became a problem. Many survivors had their clothes literally ripped from their bodies; casualties strewn about the city were often completely nude. The National Guard brought tents, temporarily alleviating the need for shelter. Providing fresh drinking water to the parched survivors turned into a huge problem; military trucks brought in water.

At the time, total damage was estimated to be in excess of $125 million. Estimates of casualties varied, ranging from 408 to about 575. The higher counts proved to be more accurate.

The war having ended a year and a half earlier, many of the farmers who had come to town to earn inflated wartime wages packed up their trucks and picked their way out of the broken city rather than rebuild. Private insurance companies eventually paid for much of the damage. Those who remained started over. Texas City was rebuilt; it stands today in spite of one of the greatest free-ranging explosions in U.S. history. The city may have been slowed in its development, but residents remember: it was one hell of a blast.

Nuclear catastrophe obliterates region of Soviet Union

SVERDLOVSK, U.S.S.R., February 1958—Invariably, researchers looking into the dynamics of significant accidental explosions are confronted with incidents that may have been nuclear in origin. The core question—whether the world has ever experienced a free-ranging, unplanned nuclear blast—is much more difficult to address than one would first suppose. Hiding such an incident would seem difficult, but perhaps it has been done.

Evidence strongly suggests that an accidental nuclear blast did occur in Russia. There may be people living who know exactly what happened, but our chances of talking with them—even in these days of glasnost—are about as good as buying fresh pork in Riyadh, Saudi Arabia.

In the early '60s, a Soviet emigrant wrote, "We crossed a strange, uninhabited, unfarmed area." (Travel by automobile in the Soviet Union was very unusual then, but apparently this obscure person was an exception to the rule.) "There were highway signs for the next twenty to thirty kilometers warning drivers not to stop," he reported. "The land here was completely empty—no villages, no towns or cities, no people or farm animals, and no cultivated land. Only stone chimneys remained where houses once stood. The driver proceeded through as quickly as possible. He insisted that all windows remain tightly closed and there be no slowing or stopping for any kind of rest or relief."

The area in question lies more than a thousand miles inside the Soviet border in the Ural Mountains, near the city of Sverdlovsk. Soviets refer to it as the Chelyabinsk region. To locate the exact region on a map, look about 800 miles north of the Aral Sea and 600 miles east of Kazan.

As early as July 1959, reports leaked out to the United States, England, and Germany, that the city of Kyshtym, south of Sverdlovsk, had been obliterated by a nuclear disaster. Whether this was the result of an actual nuclear blast, as some travelers and immigrants reported, is still—twenty years later—not known with certainty. Eyewitness accounts often conflict, and they allude to conclusions that perhaps are irrational.

The basic facts have unraveled as follows.

Initial published accounts of the incident were so obscure that most Western journalists missed them. Late in 1959, a tiny article appeared in a German-language newspaper published in Argentina. According to the article, Soviet doctors had been dispatched to Sverdlovsk with great haste to assist in the treatment of workers hurt in a nuclear accident. The article claimed the nuclear accident had occurred in February, 1958, and that significant numbers of villagers suffered from severe radiation burns.

This account—an East German physicist's analysis of what he had been told about the incident—concluded that the relatively catastrophic disaster was not the result of a single nuclear event. The physicist wrote that a gradual breakdown of a filtering system had caused fugitive dust from the numerous nuclear weapons plants in the region to spread over several thousand square miles, rendering them uninhabitable. (It is difficult to envision a circumstance wherein nuclear dust hot enough to contaminate a large section of countryside could have been manufactured by choice—much less by accident.)

Later, satellite reconnaissance confirmed that two huge sausage-shaped pieces of real estate were contaminated by high levels of radiation. These areas presumably could have been "dosed" by prevailing winds, but it stretches the imagination a bit to accept this analysis.

Representatives of the U.S. government, led by Vice President Richard Nixon, visited Sverdlovsk in July of 1959. The group was part of a goodwill entourage intent on fostering a bit more openness relative to nuclear weapons in the hope of precipitating an eventual test-ban treaty.

Hyman Rickover, founder of the U.S. nuclear Navy, was a leading member of that delegation. Some members carried dosimeters, and they reported their readings to Rickover, who acted as the group coordinator of such information. As a result of this visit to Sverdlovsk, Rickover became convinced that there had been "some sort of atomic explosion" in the region.

U2 pilot Gary Powers tried to film the region for the CIA in 1960 but was shot down by the latest and most effective Soviet missiles guarding the area. One can reasonably question the Soviet decision to place their newest and best missiles a thousand miles from any border on a speck of land with little global significance. U.S. intelligence services failed, as far as we know, to secure photos of the region until much later, with the advent of high-resolution photo reconnaissance satellites.

Other on-the-ground reports may have come to the CIA, but if so, they are still classified and thus unavailable to people interested in explosions. Many unofficial reports are available, however, mostly courtesy of Jews who emigrated to Israel.

One emigrant writing in the mid-70s recalled being driven through an area 140 miles south of Sverdlovsk, where drivers refused to stop, slow down, or even open the car windows. Government personnel drove passengers over these roads at relatively high speeds, minimizing time spent in the region. Collectively, this and other "car stories" confirm the existence of this broad, uninhabited region near Sverdlovsk.

Another emigrant, Mikhail Antonovich Klochko, sought political asylum from the Canadian government in 1961 at age fifty-nine. Klochko had been a director of the Nuclear Chemistry Laboratory in Moscow. Canadian authorities persuaded him to write out long, detailed accounts of his experiences with Soviet nuclear programs. These reports enumerated instances of callousness and carelessness with radioactive materials. They also confirmed that the Sverdlovsk region contained a nuclear weapons manufacturing complex. But Klochko had no firsthand knowledge of any accident south of Sverdlovsk. He could only retell rumors and gossip.

In 1976, an exiled Soviet biochemist, Zhores Medvedev, provided one of the first—yet still not firsthand—confirmations of a nuclear catastrophe. He said he had heard that scientists had improperly stored highly radioactive nuclear waste in the region for many years. According to Medvedev, the material eventually reached critical mass, overheated, and exploded with an enormous blast, throwing radioactive dust about the countryside. Although it does not seem logical to scientists that nuclear waste could explode under such a circumstance, Medvedev said that thousands were affected and hundreds died.

As a result of these statements, antinuclear groups attempted to force the U.S. government to release intelligence reports and analyses on the Sverdlovsk incident. U.S. intelligence authorities were understandably wary of these requests, but eventually, they did produce some ten or twelve highly edited reports.

Again, not one of the reports was filed by an actual eyewitness. Although most nuclear experts maintain that it is virtually impossible for nuclear waste to explode, about half the accounts mention a terrific explosion. Accounts also describe the evacuation of residents from the region and report that previously occupied cities were torched or otherwise leveled. All livestock apparently was shot—a most drastic measure in meat-poor Russia.

In 1978, Zhores Medvedev traveled to Oak Ridge, Tennessee, where he met with American nuclear experts to discuss the incident. Reports indicate that the Americans were initially skeptical but did agree to do additional research. Based on commonly published Soviet medical records, the experts eventually concluded that a major nuclear contamination did occur in 1958. Medvedev went an additional—and perhaps unwarranted—step, concluding that some sort of nuclear explosion had occurred, but this hypothesis was hotly debated among U.S. scientists. While agreeing that the area was contaminated, they felt a spontaneous nuclear explosion generated by improperly handled atomic waste was impossible. The answer may lie in the popular theory that an accidental conventional explosion threw nuclear debris about the region.

Without Soviet corroboration, which appears less and less likely as the years pass, we will only know a few things for sure: a great area in the vicinity of known nuclear weapons manufacturing and storage facilities was devastated by high levels of radiation; massive numbers of Russian peasants were removed forcibly from the region; and more than thirty Russian villages were dropped mysteriously from official maps.

Why the region became dangerously radioactive is still uncertain. Current satellite photos show little more than desolation. Since no outside Russian or Westerner examined the supposed impact area until several years after the incident, we cannot be sure whether a blast did or did not occur.

If a nuclear blast did occur in the region, it was certainly an accidental one. (Even callous Soviet physicists do not knowingly contaminate thousands of square miles of country.) Perhaps no other country is capable of insulating its mistakes so effectively, making verification an impossibly difficult chore.

This nuclear accident—if one ever actually occurred—may be the only example of a free-roaming nuclear blast in the history of the world.

EDITOR'S NOTE: In September 1989, The Soviet newspaper *Red Star*, a traditionally conservative organ of the Soviet Defense Ministry, revealed that thirty-five years earlier the Soviet military had dropped an atomic bomb close to its own troops in the southern Ural Mountains. The "exercise," undertaken on September 14, 1954, was designed to test troops' ability to fight in a region contaminated by radiation, the article said.

According to the report, no fatalities or injuries were recorded at the time of the test; however, "long-term effects of the radiation were never taken into account." The blast "eliminated all landmarks on the terrain and the area become unrecognizable," the paper stated. The article described terrified young soldiers taking cover from the blast in foxholes and behind low mounds of dirt and said that the heat of the explosion was so great that it melted tanks "and soon everything was covered with stones, dirt, and dead animals."

References in the article to the effect of the bomb on troops and the exercises that actually took place after the blast were vague. Some foxholes were covered and other shelters had double doors, the report stated. Soldiers were issued gas masks and uniforms to help ward off the radiation, and medical centers were set up in the area. Soviet officials reportedly attempted to limit radioactive fallout by waiting for ideal weather conditions and by dropping the bomb from a plane and detonating it at a low altitude of between three hundred and five hundred yards. Scientists said these efforts were successful in reducing radiation, the article reported.

Perhaps as glasnost continues to revolutionize the Soviet press, the details of these and similar incidents will emerge from the shroud of secrecy under which they have been kept for so many years.

Fire touches off explosive-laden truck in downtown Roseburg; blast levels 7 city blocks

ROSEBURG, OREGON, August 1959—It isn't always the net tons of explosive that put a great blast in the record books. Under some circumstances, position and timing count heavily.

Oregon in the late '50s was a brawny, tough, entrepreneurial place, populated for the most part by farmers, ranchers, lumberjacks, and small businessmen. Hardware stores still stocked and sold caps, fuze, and dynamite by the individual piece for those who could not buy by the case.

Roseburg, Oregon, lies in the south third of the wide, fertile Umpqua Valley between the western coastal mountains and the high Cascades to the east. In 1959, approximately nine thousand citizens populated the little rural community some two hundred miles north of the California state line.

George Rutherford pulled his Pacific Powder Company ten wheeler into Roseburg about 9:30 P.M. on Thursday, August 6, 1959. George knew when he left the explosives factory at Tenino, Washington, that he would get into Roseburg too late to offload and start back. He parked his rig near Gerretsen Building Supply in anticipation of delivering his cargo of explosives at the full-line hardware store early the next morning.

Rutherford locked the box and cab of his nearly new '58 Ford truck and walked three short blocks to the Umpqua Hotel. Curiously, given his dangerous cargo, he almost seemed to have staying in the hotel that night as much on his mind as making an efficient delivery.

Gerretsen Hardware kept all but small quantities of powder—which they mostly sold by the stick—five miles out of town in a powder magazine. Larger, ton-lot sales from the magazine were made on a wholesale basis to area ranchers, loggers, and miners. This was not Rutherford's first delivery to Gerretsen's, so he undoubtedly knew the material would not be offloaded downtown at the retail store where he had parked.

The truck he so cavalierly shoehorned into the

ROSEBURG, OR, Aug. 7—FIRE RAGES THROUGH ROSEBURG—Shown on the previous page are the smoldering remains of Roseburg following the fire caused by the explosion at a local lumber center. The fire, after raging out of control, was contained in an area of five or six blocks. (Douglas County Museum photograph.)

parking spot in the middle of town and walked away from was actually loaded to the gills with high explosives. Ammonium nitrate is a basic agricultural fertilizer; it is not classified as a high explosive or thought of as particularly dangerous by knowledgeable people. Rather than being straight agricultural fertilizer, however, this load actually contained four-and-a-half tons of nitro carbonitrate, or Car-Prill—prilled ammonium nitrate treated with diesel fuel and mixed with ground nutshells—and four thousand pounds of straight dynamite packed into fifty-pound boxes.

As significant blasts go, the ingredients for a whopper seemingly were not in place. Thirteen thousand pounds of low-yielding, slow explosives is not a great deal—unless, for some reason, the material is sitting right in the heart of a small town.

Marilyn Tandy picked up her husband Dennis at the Nordic Veneer Company thirty minutes after the shift change at midnight. (Working late into the swing shift that night was a fateful piece of business in Dennis Tandy's case.) On their way home through the middle of town, they passed Gerretsen Building Supply. Dennis spotted a small blaze in some trash cans behind the store and got out of the car to see if he could pull some of the fuel away from the blaze in an attempt to contain it. On his instructions, Marilyn drove to a service station a block away, where she turned in the alarm.

Marilyn got back to Gerretsen's at about the same time the prompt small-town fire truck arrived. She noted that the blaze was no longer small. Great clouds of black, greasy smoke blanketed the area. The heat being generated was sufficient to make it uncomfortable to remain in the area. Mrs. Tandy backed the car down the block about one hundred feet to an empty lot and parked.

In the absence of the fire chief, who was in the hospital with a heart condition, Lyle Wescott, the assistant, snaked off the hose and attached it to the nearest main. It would have been totally impossible for one man to drag that hose, even under conditions of extreme emergency. Fortunately, Richard Knight, a young service station attendant from up the block, ran down to give Wescott a hand. Knight had received training as a fire fighter in the U.S. Air Force.

A Gerretsen delivery truck blocked the drive, limiting access to the explosives truck and the fire. Young Tandy went to work with diligence, determined to break into the vehicle with the fire-truck axe. That accomplished, the three were able to work their way close enough to the hot, smoky fire to read the signs on the two-and-a-half-ton Ford. Horror of horrors, twelve-inch signs in big red letters on all four sides warned that the truck contained explosives. It was 1:13 A.M.; the three men had one minute to live, but they remained at their stations.

At 1:14, Marilyn Tandy's car was blown across the lot about one hundred feet. It landed twisted, bent, and battered, but upright. Chuck Lynch, a passing motorist, managed to execute an emergency U-turn away from the epicenter just at the moment the explosives went up. The force belted his car down the street at seventy miles an hour. Lynch would have wrecked horribly, but the same concussion bounced, slowing him down as if he had hit a wall of Jello. The car's gaudy exterior chrome was all torn away, but Lynch was unhurt.

Carol Merical, a waitress at the Umpqua Hotel, was walking home at the time the blast hit. She was blown more than 150 feet down the street. All her clothes, her glasses, and even her shoes were ripped off. She remembers that the pavement felt hot to her bare feet afterwards, but it may only have been the warm summer evening. Carol's close friend Donnie Berg walked a few steps behind her when the truck blew. Donnie was

ROSEBURG, OR, Aug. 7—DEVASTATION SCENE—The huge crater left by the explosion that demolished several city blocks and damaged others extensively. A truck loaded with explosive material blew up at Gerretsen Building Supply Co. in downtown Roseburg, wiping out surrounding buildings. More than a dozen persons were killed and dozens of others hurt. (Douglas County Museum photograph.)

ROSEBURG, OR, Aug. 8—BLAST WRECKS SCHOOL—Central Junior High School of Roseburg looked like this after the blast, which trashed more than twenty city blocks. In the foreground is the crater made by the explosion. (Douglas County Museum photograph.)

ROSEBURG, OR, Aug. 8—FORCE OF THE EXPLOSION—Gov. Mark Hatfield and a group of officials inspect the ruins following the Roseburg blast. Roseburg Police Chief Vern Murdock and Mayor Arlo Jacklin are at Hatfield's left (second row). Behind them is the twisted wreckage of a train devastated by the explosion. (Douglas County Museum photograph.)

ROSEBURG, OR, Aug. 7—BUILDING GUTTED BY BLAST—The skeleton of an automobile dealership showroom near the center of the blast. A man was thrown through the window of this building by the force of the explosion and survived to tell the tale. (Douglas County Museum photograph.)

thrown into the remains of the Stock Motors Co. lobby, where she died.

A total of thirteen people died; 125 received injuries of varying seriousness. All of the fire fighters on the scene became instant casualties. Driver Rutherford was bounced out of his bed at the hotel, which itself suffered extensive damage. He spent two days in the Roseburg hospital before returning to his home in Chehalis, Washington.

Although the term was not in vogue in that era, the blast committed an ultimate act of "urban renewal." Seven city blocks were leveled in the heart of Roseburg and another twenty-eight were trashed to some extent. Most windows within a four-mile radius were knocked out; some as far as nine miles out were smashed. Area residents as far as seventeen miles away reported feeling the blast but said they assumed it was a sonic boom from one of the new jet fighters that increasingly were flown over the area.

Downtown, the blast bit a crater forty feet in diameter and fifteen feet deep in the place where the hardware store once stood. Heavy rail cars resting on a siding across from the store were picked up, bent, and then dumped on the other side of the tracks in a crazy, twisted pattern.

Some of the residents of Roseburg must have eaten live toads for breakfast, they were so tough. Fred Siles, a mill worker, was blown through a huge automobile agency showroom window. He got up, dusted himself off, responded to waitress Merical's calls for help, found a functional car, and drove the bleeding, battered, woman to the hospital.

At the city hospital, chaos was in season. There were not enough nurses and doctors to handle the load. Calls for additional emergency helpers and supplies were futile. Most phone and power lines were chopped out by the blast. Fires continued to burn through the rubble the following day, destroying what had not been crushed by the original blast. Dust and smoke hung over the region like a thick mantle.

Many Roseburg merchants who had worked all their lives to build their business now owned unruly piles of rubble. Insurance money eventually provided a new start, but it was a long time coming. Twelve million dollars. That was the price tag the insurance company and the merchants placed on the damage to the heart of the city. (Today the bill would have been at least triple that amount.)

Driver Rutherford went back to his home and job in Washington state. He continued to work for Pacific Powder, but not as a truck driver. The explosives company sold all its trucks and put Rutherford in the production department as a laborer.

Officials in Oregon complained about the fact that it was so "criminally" easy to drive a truck loaded with explosives into a community and then temporarily abandon it to whatever fate provided. Others pointed out that the driver had violated at least two Interstate Commerce Commission (ICC) regulations regarding transport of explosives.

Authorities said he had violated the rules requiring that explosives-laden trucks be taken straight to their unloading point and never left unattended. He had also disregarded the stipulation that populated areas must be avoided if at all possible.

Additional rules were proposed following the incident, including one mandating that when a delivery of explosives is made, all adjacent areas be promptly evacuated and considered unsafe. It was also recommended that fire-department personnel be given additional training in the proper methods of fighting explosives fires.

Rebuilding, to a great extent, was hampered by Roseburg's city fathers who insisted that any urban

ROSEBURG, OR, Aug. 7—ROSEBURG BUSINESSES RUINED—Bill Stock Motors Company had more than wrecked cars to repair after the explosion that ripped through Roseburg, leaving Oak Street in shambles. (Douglas County Museum photograph.)

ROSEBURG, OR, Aug. 7—BUILDING COLLAPSES FROM FORCE OF EXPLOSION—The view looking west on Oak Street, which was virtually obliterated by smoke and flames as a result of the giant explosion-caused fire. (Douglas County Museum photograph.)

renewal be done with the blessing and approval of the city renewal committee. Fighting over the plan and the long interval until the insurance companies paid up worked together to keep anything of substance from happening for several years.

Like the real estate, the people of Roseburg persevered and eventually recovered. Today, casual tourists passing through town seldom realize that little Roseburg, Oregon, endured one of the largest explosions ever to run free upon an "our town"-type community.

Shipload of munitions explodes in Havana Harbor; Castro points finger at U.S.

HAVANA HARBOR, CUBA, March 1960—At times, history seems to repeat itself in an uncanny sort of way. It would, for instance, seem improbable that two ships should mysteriously and disastrously explode, years apart, in basically the same place in the same harbor, and that both explosions should act as catalysts for wars. But the improbable happened. The historic explosion of the *Maine* led to a shooting war, while the other, sixty-two years later almost to the day, contributed to the Cold War in which the superpowers are still engaged.

Early on the morning of Friday, March 4, 1960, the 4,300-ton French coastal freighter *La Coubre* took advantage of a flooding tide to slip up into Havana Bay, past Morro Castle, which guarded the harbor's mouth, to a downtown pier-side berth.

The arrival of *La Coubre* in Havana culminated what had been an extensive diplomatic tug of war, pitting the Kennedy brothers in the United States against Fidel Castro in Cuba. As a result of the imbroglio, relations between the two nations were going to the tub rapidly.

Castro wanted to buy arms for his fledgling communist government, while the United States was doing all it could to prevent the island from becoming an armed camp. American teams followed the arms-shopping Cubans around Europe trying to persuade its friends and allies not to sell to Fidel or extend credit for any possible purchases.

Late that January, Castro's agents finally persuaded a Belgian munitions broker in Liege to sell them a mixed load of seventy-six tons of small-arms ammunition, mortar rounds, artillery shells, and grenades. In the total scheme of things, the shipment was very small. Most of the munitions were of a size and type that would function in American-made weapons left behind from the Batista regime.

At this time, fifteen months after Castro's rise to power, relations between the two nations were still

HAVANA, CUBA, Mar. 14—WORKERS EXAMINE WRECKAGE—On the previous page, workers taking part in salvage operations examine a large piece of wreckage. Warehouses and homes for blocks around the blast site were heavily damaged by large pieces of flying metal. (Photo courtesy of AP/Wide World Photos.)

somewhat amicable, but they were souring quickly. It would have been far easier, quicker, and cheaper to bring the munitions in from Miami, just ninety miles away, rather than ship them directly from Belgium, but making it easy on the Castro government was not what President Kennedy had in mind.

Don Chapman, an American from North Bend, Nebraska, was aboard *La Coubre* as it passed through the narrow channel into the seedy, run-down waterfront area of Havana to berth. The vessel tied up on the northwest side of the harbor. (Later, Cuban intelligence police would nab Chapman, holding him for two full days of interrogation before satisfying themselves that he had nothing to do with the disaster.)

Chapman stood idly by on the vessel's bridge, watching the swarm of ant-like Cuban men below unload with cargo booms and nets. By 3:00 P.M. on March 4, the husky, sweating longshoremen had made a respectable dent in their workload. The stinking, oily harbor, combined with the humidity and heat, made remaining on the bridge something of a chore (but watching the drama that would soon unfold more than compensated). Even Chairman Castro himself circled above in his helicopter, chortling over his success in sticking his thumb in the dreaded Yanqui eye.

Suddenly, a sharp, gutsy explosion in the afterhold shocked the vessel severely. Chapman, apparently on something of an alert due to the nature of the cargo, immediately raced down through two decks to relative safety on the pier. Whistling, snapping shells filled the air, threatening his passage, he later recalled.

Initially, the explosion caused the loss of the entire aftersection, creating a sharp list to port. In a few hours the vessel rolled over, taking the remaining cargo and crew with it. Fire raced through the ship, which began taking water through gaping holes in the plates. Burning oil filled the immediate area, flaming the creosote-soaked pilings and the pier.

Gabrielo Delgado, a Cuban roustabout, attempted to fight the fire in the rear hold. He reported later that the scene was something out of Dante's *Inferno*. Delgado said he faced a wall of flame as exploding bullets and shrapnel flew past his head, careening off the steel bulkheads. Miraculously, he was not injured, but many others were not so fortunate.

A second, more severe blast threw bodies and pieces of bodies—mixed with wooden crating and pieces of steel plate—about the hold. Delgado and the other survivors retreated to the deck, happy for the life they still possessed.

On the pier, the scene was quickly becoming a battlefield. Fidel's crew in the chopper had been able to radio for help immediately after the first blast. Ambulances from downtown Havana responded with great dispatch. Some of these emergency vehicles were trashed by subsequent blasts, however, leaving their cargo of wounded in even more desperate straits.

The exploding mortar rounds set off other blasts in rear warehouses. Before the situation stabilized, a total of three large storage warehouses packed with munitions disappeared off the face of the earth. For blocks around, windows were shattered. Debris rained down on cars parked along the narrow streets. Ancient cement balconies overhanging the street—relics from the pre-1898 Spanish era—were damaged beyond safe, productive use. A hapless office building that stood in the blast zone west of the warehouses was damaged so badly that it was eventually loaded into dump trucks and hauled away. The entire waterfront district went instantly from shabby to more shabby as a result of the continuing blasts, many of them from rounds designed to crack open cement bunkers. Buildings for miles

HAVANA, CUBA, Mar. 14—TWISTED REMAINS SALVAGED—In his haste to get seventy-six tons of munitions unloaded from the French ship SS *La Coubre*, Fidel Castro lost a vessel, the pier to which it was tied, three warehouses, and an office building. A total of seventy-seven people died and more than one hundred were hospitalized. Here, a giant derrick clears the ship in the aftermath of the blast. (Photo courtesy of AP/Wide World Photos.)

around the *La Coubre* were bent and twisted. The pier at which *La Coubre* docked was a total loss.

Customarily, munitions ships are anchored out in the harbor at least one mile from anything expensive. From this safe distance, they are unloaded by lighter ships that carry their cargoes ashore. On shore, hazardous-materials discipline is maintained, with all the beached goods being cleared away before new materials arrive.

But, in his haste to get the cargo off, Fidel lost a vessel, the pier to which it was tied, three warehouses, and an office building, not to mention a total of seventy-seven lives. Ironically, forty of these were onlookers. Only thirty-seven of the dead were stevedores. Those injured and requiring hospitalization totaled well over one hundred. Many onlookers reportedly were killed when, out of curiosity, they picked up stray hand grenades strewn about the area by the blasts.

A second vessel, apparently unloading other nonmilitary cargo, was tied up at the same pier as the *La Coubre.* It managed to clear the area, however, escaping any significant damage.

Blame for the blast was assigned to the Americans. Fidel, in one of his typically fiery speeches, said, "Officials of the U.S. government were interested in keeping arms from Cuba. Since the North American consul and military attaché in Belgium has tried by every diplomatic means to keep Cuba from making this purchase, we have the right to think that the Americans have tried to prevent delivery by other means."

Cuban newspapers picked up the rhetoric, calling in the most strident terms for vigilance in regard to an overt U.S. invasion of Cuba that was "expected" any day. A linkage was made between alleged American attacks by planes loaded with "counterrevolutionaries" and the current U.S. push to cut Cuba's sugar quotas.

(At the time, the United States was still paying Cuba sixty-five cents a pound for sugar that cost about three cents a pound on world markets.)

American reporters who were allowed to visit Cuba attempted to discover how the blast had occurred. They reported that numerous dockside laborers were smoking on the job and that matches and cigarettes were not confiscated from those handling the explosives. At a meeting of the dockworkers union, survivors claimed that a defective cargo net gave way under the load, dumping sensitive mortar rounds into the rear hold, where they detonated. The dockworkers readily admitted that they did not know the proper methods of handling munitions.

Pressing his propaganda opening among the Cubans, Castro took two cases of hand grenades up in his helicopter and threw them out from six hundred feet without noticeable effect. Escalating the rhetoric a bit, he claimed that his demonstration proved that the blast could not have been an accidental one caused by a torn cargo net.

Without some sort of release of data by U.S. intelligence sources, it is difficult to know for sure if the United States was involved, or why Castro chose to throw hand grenades out of his chopper door in an effort to demonstrate that it was, since the boat was loaded with a variety of munitions. Perhaps it was because of the number of casualties the hand grenades caused among the onlookers.

The national hysteria surrounding the incident faded quickly on both sides. Cubans, afraid of further antagonizing the giant to the north, quietly found alternative, Soviet sources of supply. American diplomats—their hands full in Vietnam and wary of the jingoism that brought the United States into Cuba sixty-two years earlier—let the matter slide.

Whether accidental or deliberate sabotage, it was a

good, healthy blast, but one obscured by time and place and other highly newsworthy events. Seventy-six tons of pure military explosives—tamed by heavy metal cases and packing—was so small by modern comparisons that the event has faded even further in the collective memory.

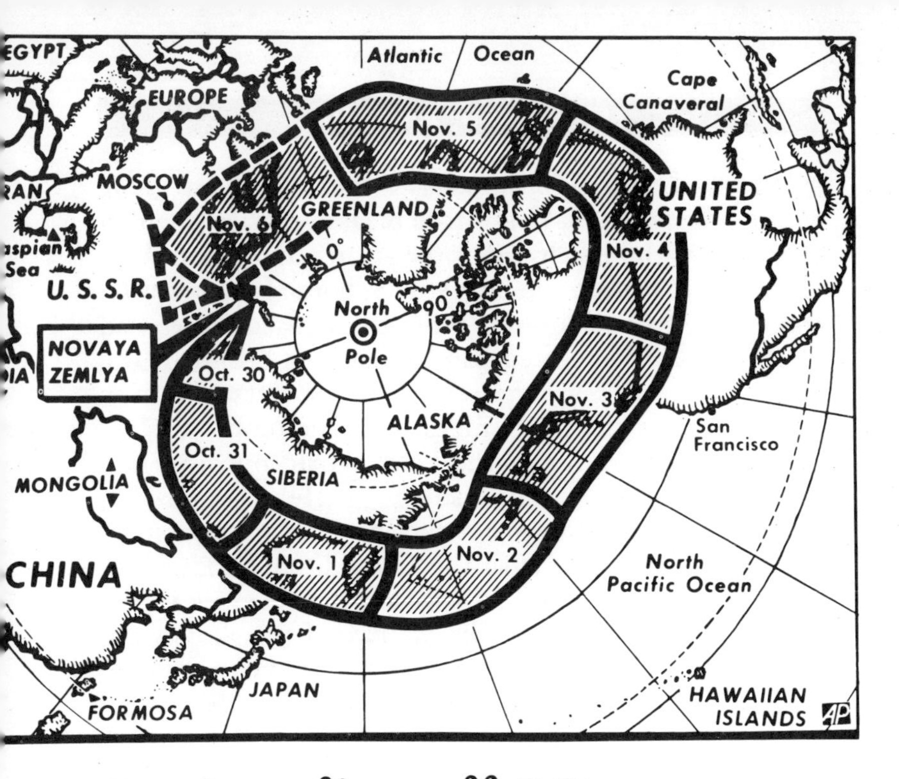

Soviets fire off 50-megaton atomic "superbomb" in Arctic

NOVAYA ZEMLYA ISLANDS, U.S.S.R., October 1961—The world accurately perceived that the Soviets were sticking a military thumb in its eye when Nikita Khrushchev announced that his government intended to resume atmospheric atomic testing in October, 1961. The terse, stiff Soviet announcement contained an incredible amount of bravado.

Not only did the Russkies intend to break their self-imposed thirty-four-month moratorium on atmospheric testing, they also boasted that they were going to fire off the largest man-made blast ever unleashed on the face of the earth. Fifty megatons, they said. The blast was to exceed the largest U.S. bomb by more than three times. On March 1, 1954, the United States had detonated a fifteen-megaton monster at Bikini in the South Pacific. This puny record, the Soviets said, was about to fall to their fifty-megaton superbomb.

For thirty-four months now, world opinion had mostly managed to contain the nuclear genie. The French were only a minor exception. French physicists, anxious to catch up with the United States, England, and the Soviet Union, had fired off four devices in the Sahara in Algeria, North Africa. Two of the shots were of such low yield they were classed as virtual duds. Other than this, there had not been any really significant shots.

Fifty megatons is equal to fifty million tons of TNT. Previously, the largest conventional TNT blasts ranged from about a million pounds in Belgium up to about eight million pounds in Halifax Harbor, Nova Scotia. In the case of the Soviet bomb, we are looking at multimillions of tons—not pounds—of nuclear explosive. Each of these long tons contains more than 2,200 pounds . . . numbers so big they defy valid comparison.

Scientists throughout the world were horrified. Many talked about the blast shaking the whole earth; blinding innocent, unaware bystanders within a two hundred-mile radius; and setting off uncontrollable

WASHINGTON, D.C., Nov. 6—HOW RUSSIAN SUPERBOMB FALLOUT CIRCLES GLOBE—The map on page 125, based on one released by the U.S. Weather Bureau, shows the path of radioactive fallout from the Soviet nuclear blast of October 30, 1961. The bureau estimated that after one week, the cloud was rolling across the North Atlantic and over Scandinavia. (Photo courtesy of AP/Wide World Photos.)

chain reactions. Some even mentioned nudging the earth off its axis; others, vast fire storms.

Much of the rhetoric was hopelessly pessimistic—full of Cold War propaganda. Khrushchev bravely responded by announcing that not only were the Soviets planning a fifty-megaton blast, but that they were actively considering a hundred-megaton encore. This much explosive power is virtually impossible for mere mortals to comprehend.

Practically speaking, atomic-bomb construction had reached an upper limit. U.S. bombs used in anger and tested by scientists had been in the kiloton range. Beyond a couple of megatons, sufficient fissionable material could not be packed into the bomb's structure. A few larger bombs had been tested, but as far as the United States was concerned, the race was for smaller, more reliable devices for use in atomic cannons and Davy Crockett-type missiles.

Reportedly, the United States had a few twenty- to twenty-four-megaton devices, but these were not considered to be particularly effective in a military sense. Pinpoint accuracy was thought to be much more important than blowing an enemy into even finer dust. Rational nuclear physicists wrestled with the law of nature that dictated that in order to double the area impacted, the size of the bomb must be squared. It took many years, but even the Soviets eventually came around to this mechanical/philosophical realization.

Although the West has never discovered the exact structure of the Soviet bomb, the basic design from which it sprung was common knowledge. For double-digit-yield nuclear devices, nuclear scientists resorted to a three-stage concept known to bomb builders for many years.

The first stage is a nuclear-fission trigger constructed of enriched U-235. The second stage, or fusion stage, is made up of lithium and deuterium. (Deuterium, a double-weight hydrogen, combines with lithium to form an opaque solid; it is the component that gives the H-bomb its name.) When triggered, the atomic bomb acts as a giant blasting cap, detonating the lithium/deuterium and initiating the fusion reaction. This fusion releases a great blast pressure, including copious amounts of heat. Combined with large numbers of loose neutrons, this powerful second stage causes the outer shell to undergo a fission reaction. The outer fission jacket, or third stage, is comprised of common uranium U-238, which is relatively inexpensive and easy to refine and mill.

Technically, there is no limit to the size of bomb that can be produced using a three-stage mechanism of this nature. When Khrushchev claimed that the Soviets would detonate a fifty-megaton device and that a hundred-megaton monster was under consideration, no one seriously doubted that such a bomb could be built or that it would work as promised. In fact, scientists throughout the world took Khrushchev's boasting very seriously. Everyone agreed that there was no upper limit; only good common sense could dictate the size of the bomb men could build. Good common sense seemed—in many regards—to be in short supply in the Soviet Union in 1961.

Some physicists even went so far as to calculate that the blast would release 2,700 pounds of radioactive material into the atmosphere. They calculated that the fireball from the device would have a radius of 3.5 miles, either necessitating a very high-altitude shot or bringing about extremely dirty results as the blast scraped up material from the earth and carried it into the stratosphere as highly radioactive particulate matter.

Chemist, Nobel Prize winner, and outspoken nuclear critic Dr. Linus Pauling stated categorically that firing the device would cause forty thousand new cases of cancer within a few short years and that forty thousand deformed babies would be born in the years to come. He also said that everyone within two hundred miles would suffer permanent retinal damage. Statements of this nature are difficult to refute because they can be dropped with such impunity.

Newsweek, reacting in absolute terror, published a map of Manhattan Island with concentric lines graphically depicting the extent of damage that a fifty-megaton nuclear device would do—theoretically—to New York City. The report claimed that the blast would dig a hole three hundred feet deep and one mile wide. Everyone within three-and-a-half miles would be killed. Even out to thirty-five miles, it was estimated, everyone would receive second-degree burns from the fires that would devastate the region.

Initially, there was a great deal of confusion regarding the actual planned location of the test shot. Western analysts universally assumed it would be in Central Asia near Semipalatinsk, which is east and north of Lake Balkhash in Soviet Asia. Hypothetical fallout maps depicted a radioactive band of material deposited over northern Japan, Sakhalin Island (U.S.S.R.), and on a line roughly separating the United States from Canada.

When all was said and done, the Soviets actually tested their bomb at their range in the Arctic. It was launched from the Novaya Zemlya Islands, between the Barents and Kara Seas, and detonated in the stratosphere.

Perhaps because of extensive protests on the part of Norway, Sweden, Denmark, Canada, and Finland, the Soviets detonated the device at a very high altitude. Both American and Soviet scientists monitoring the blast reckoned it to be in the fifty-seven megaton range. It therefore exceeded its original maker's estimate in yield by about 14 percent. Perhaps

Khrushchev instructed his bomb's makers to over-build to avoid possible embarrassment.

The blast turned out to be 2,600 times as powerful as the one that devastated Hiroshima. Up to the moment of detonation, the equivalent of about 170 megatons of nuclear explosives had been detonated by France, Britain, the United States, and the Soviet Union combined. On that Tuesday morning, the combined total of nuclear explosives expended worldwide increased by almost a third in one fell swoop.

No reports have been brought out of the Soviet Union by emigrants regarding reported or actual blindings within the two hundred miles of the test site. (Apparently some small increases in worldwide background radiation resulted, but they proved to be so insignificant that even rabid opponents of nuclear power soon forgot about them.)

Perhaps the dense cloud cover over the region, common that time of year, protected humans within range of the shot. The fact that the region was so sparsely populated certainly contributed mightily to the lack of reports about either the blast or any resulting casualties.

As is so characteristic of projections involving nuclear power, most of the prophesies proved to be little more than factoids with no basis in science or common sense. It was, however, as the Soviets promised, the biggest bomb to date. It seemed to accomplish the objectives set forth by Khrushchev and his cronies (i.e., terrorizing the nations of the earth). It also masked the significant shortcomings of Soviet delivery systems at that time, which in a full-blown exchange would have caused the U.S.S.R. to be annihilated.

A hundred-megaton device was never tested by either side. Khrushchev said he was concerned about a blast that could largely damage the Soviet Union itself. In response, the United States elected to make its bombs smaller and cleaner. We concentrated on problems related to increasing accuracy and weight-to-yield ratios.

Although the superbomb set a record that probably will not be broken for many years, the results for the Soviets were not particularly beneficial. The United States used the occasion to launch a massive—and largely successful—propaganda campaign against the Soviets.

The incident itself basically dropped into the cracks between salvos of propaganda and was soon forgotten by a world that had already come to tolerate smaller underground tests.

Ice show no holiday in Indianapolis; blast kills 64, injures 385

INDIANAPOLIS, INDIANA, October 1963—Because it is impossible to see vapors of explosive gas, the study of great historic explosions is somewhat abstract. Some people work with dynamite of various grades, ammonium nitrate, and perhaps primer cord and caps. Few will ever get a chance to see or play with blocks of C-4 plastique, artillery shells, or TNT.

On the other hand, virtually everyone has seen a popcorn machine. Not a terribly ominous-looking instrument of destruction, yet a popcorn machine blew up on Halloween night, Wednesday, October 31, 1963, in Indianapolis, Indiana, killing sixty-four people and injuring another four hundred or more.

It was Shriners' night at Holiday on Ice, with $3.50 tickets being offered on a two-for-one basis. More than 4,320 people took advantage of the half-price promotion and gathered at the Indianapolis State Fairground Coliseum. Parking in the lot just off Route 37 was snug, but not overflowing; it was just the kind of crowd that makes a family outing fun.

Most of the box seats in front of the cement wall separating the concessions from the arena concourse were filled. The ringside seats, especially the folding chairs down front by the ice, were also filled. Behind the cement wall, the timer was running on a popcorn machine that was connected improperly to an LP gas cylinder.

The new, fully-charged, twenty-four-gallon gas cylinder that was to be used to power the popcorn maker had been delivered that morning. Other than a common 110-volt electrical connection, the unit was self-contained. In all probability, some dollar-an-hour minimum wager threaded the connection improperly; gas fittings have reverse threads, a fact generally unknown to untrained people.

A Dixieland band belted out the finale, to the delight of audience and performers alike. For the Holiday on Ice Show's first night in Indianapolis, the

INDIANAPOLIS, IN, Nov. 2—DEBRIS BENEATH STANDS—The scene on the previous page shows the rubble below the stands that spilled the spectators watching the finale of the Holiday on Ice Show. The ripped concrete at right provides evidence of the enormity of the explosion. (Photo courtesy of Indianapolis Newspapers, Inc.)

Mardi Gras number that always brought the house down was an ironic omen. In the hall, heavy LP gas seeped out of the cylinder and ran like water along the floor of the coliseum. Apparently, no one detected its distinct odor.

Perhaps a cigarette or a sparking electrical fixture touched the gas off, but suddenly a huge, thirty-foot ball of flame burst through the concrete wall, blowing it to pieces. Chunks of concrete, some weighing up to several tons, shot out and flew through the first-class section and over to the bleacher section covered with folding chairs. The blast had enough character and intensity to throw people along with it.

What an instant before had been an orderly chorus line of glitzy, barelegged beauties was now chaos. Shredded programs, popcorn boxes, popcorn, and chunks of steel and cement—some up to twelve feet long—littered the ice. People, including a great number of children, were strewn about like so many rag dolls.

Believe it or not, one of the matrons in the audience shrieked, "It's part of the show, it just has to be!" The eighteen-piece band, not wanting to believe what lay in front of them, continued playing rickety-tickety Dixieland long after the blast. When the reality of the situation finally registered, they fled in horror, along with the many other survivors.

As is typical with gas explosions, the force was not terribly shattering. Instead, this one acted more like a giant, uncontrollable hand wafting huge chunks of concrete fifty feet through the air over the coliseum's best seats into the "next best" section of folding chairs. Dozens of people were trapped under the debris, and others suffered broken, torn, and bruised shins as they negotiated through it to safety. Survivors suffered grievously as the result of being pitched through the air onto spectators on the other side of the ice.

The blast dazed many of the show's attendees.

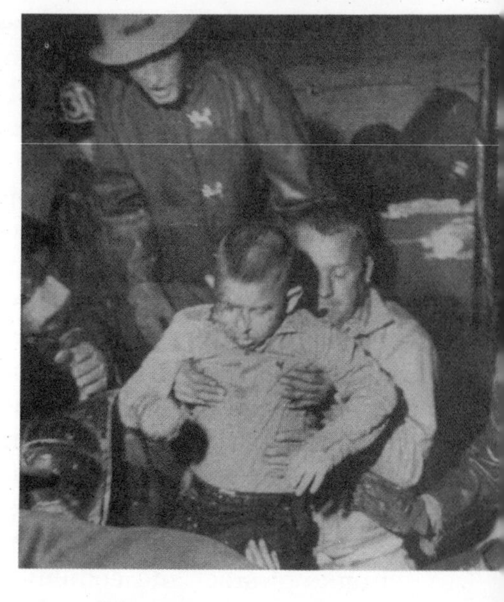

INDIANAPOLIS, IN, Oct. 31 —BOY PULLED FROM WRECKAGE—Rescue workers lift an unidentified victim from the rubble after seats and concrete collapsed on spectators watching the last act of an ice show. (Photo courtesy of Indianapolis Newspapers, Inc.)

INDIANAPOLIS, IN, Nov. 2 —DETERMINING CAUSE OF BLAST—Officials study the area where the blast originated to determine its exact cause. The disaster is believed to have been caused by the combination of a gas leak and a spark from a popcorn popper that was wired improperly. (Photo courtesy of Indianapolis Newspapers, Inc.)

INDIANAPOLIS, IN, Oct. 31—DEAD LITTER ICE—A police officer helps people search for relatives among victims lying on the ice following the explosion at the Indianapolis State Fairgrounds coliseum. Sixty-four people were killed and approximately four hundred injured in the blast. (Photo courtesy of Indianapolis Newspapers, Inc.)

Bleeding and confused, they wandered around or just lay there, not knowing what to do or where to go. Some of the survivors searched fruitlessly for family members lost in the confusion and debris.

As is true in many emergency situations, unlikely unsung heroes suddenly stepped forward. Average attendees, dressed in their evening finery just a few minutes earlier, worked feverishly to answer calls of help, performing superhumanly to lift chunks of concrete from victims. Many victims with cruelly shattered limbs had to be carried individually from the coliseum.

INDIANAPOLIS, IN, Oct. 31—WRECKAGE CLEARED—Large cranes work to clear the rubble and free those buried in the blast. (Photo courtesy of Indianapolis Newspapers, Inc.)

Naturally helpful Hoosiers descended on the coliseum en masse to help. Two huge cranes were brought in to lift the larger slabs, revealing more casualties. Fatalities among the wounded were kept to a minimum by the large, well-staffed Indianapolis medical facilities.

As fortune would have it, the National Federation of Licensed Practical Nurses was meeting in Indianapolis at the time. They hastily adjourned their convention and reported for duty. The final toll was 64 dead and 385 injured. Of the injured, 176 were classed as critical, requiring hospitalization. Mayor Albert H. Losche ordered all flags flown at half-staff.

The incident was—and still remains—the worst single catastrophe in the history of Indianapolis. Odds are that it was the first and last time a city was forced to react so frantically to an exploding popcorn machine.

Explosion at gun club rocks town

MOSCOW, IDAHO, December 1969—Mrs. Beryl Jorstad knew something was going wrong. She was unsure what exactly was stored five hundred yards from her house in the large silver trailer across Highway 8. Whatever was in it, she and her mother watched intently out of the large living room window as copious amounts of wispy, steamy-white smoke poured out of the trailer. At 1:26 P.M., the two women saw workmen around the trailer run desperately up the gentle landing toward the highway. The men made it far enough away so that none were injured when the twenty thousand pounds of ammonium nitrate stored inside went off with a thunderous roar at 1:29 P.M.

The blast sent showers of glass across Mrs. Jorstad's room as every window in the house was vaporized. As is so often true with great explosions, the concussion played odd tricks. Mrs. Jorstad and her eighty-year-old mother were not, as one would suppose, pitched across the room. The mighty blast did, however, stop the living room clock, blow away the roof, and otherwise damage the house to the extent that it took three months to repair. Flying glass inflicted extensive flesh wounds upon the pair. Oddly, their eyes were not damaged, but observers later remarked that their living room looked as though someone had butchered a hog there.

Moscow, Idaho—the second largest city of that name in the world—lies about two hundred miles south of Canada's border and some ninety miles south and east of Spokane, Washington. Were it not for a long-forgotten political accident in 1889, the University of Idaho would not have been located in Moscow. Rather than housing eight thousand students plus seven thousand permanent residents, Moscow would have evolved much as the other sleepy little logging, farming, and mining communities that characterized the region.

Highway 8 runs from the city of Moscow to Pullman, eight miles to the west. The blast was a mile closer to Pullman, but conditions were such that Moscow sustained the damage. To the south, immediately adjoining the highway, the Pullman Gun Club had purchased a small, flat piece of ground created when the previous owners crushed the rock bluffs to sell to the county as road ballast. On Sundays, club members from both communities shot trap safely into the steep basalt cliffs.

Claud Crow, an experienced powder monkey and rock crusher operator, returned from lunch in Moscow, anxious to get on with the day's work. It was a miserable, cold Tuesday afternoon. He was intent on loading drill holes and shooting the columnar basalt for his crusher. The project promised something for everyone—more space for the gun club, rock for the county, and work for Mr. Crow.

In anticipation, he had ordered twenty thousand pounds of pretreated ammonium nitrate. Conventionally, ammonium nitrate is handled as dry prilled fertilizer. It is officially classified as a blasting agent rather than an explosive. Even the pretreated material is considered to be relatively benign, noted for its safety, low price, and ease of handling.

The material was delivered in burlap-wrapped, bologna-like tubes and stored in an old eight-by-forty-foot covered van that Mr. Crow had pulled into the area directly behind the gun club building.

The crew waited till noon for the sun to warm the ice cold steel box. The oily, premixed materials were so glutinous because of the bitter cold that they would not pour into the shot holes. Before leaving for lunch, the workers fired up several kerosene salamander heaters in an attempt to thaw out the chemical mountain.

The story later circulated was that one of the heaters was defective. Ron Grotchen, a college student working part time for Crow, displayed great courage by jumping into the trailer and trying to snuff out the oily fire. When he realized his efforts were futile, he fled as well. Perhaps the heater was defective or perhaps it was improperly filled. There is no going back now to look at the evidence.

Mrs. Jorstad remembers the smoke rolling out of the trailer, the men running, and then the blast, for which she is still unable to find words, but which she says she will never forget. "The noise was louder than anything I have ever heard," she says.

The house next to hers, slightly closer to the explosion, was destroyed. Months later, wrecking crews bulldozed it into a pile and burned it. (As is true in most cases of massive disaster, insurance companies took nearly four years to finish making settlements.)

Lorraine Hudson, wife of Moscow's chief of police, chose that fateful day to wash and wax her hardwood floors. Unlike Beryl Jorstad, she felt the pressure thump as she walked to the next room. No physical damage occurred to her house, located in residential Moscow, eight miles away.

Mr. Crow and his workers were able to stop traffic on the lightly traveled Highway 8. Other than Mrs. Jorstad and her mother, whom the men were unable to warn, no one was hurt. A mobile home up a draw near the Pullman clubhouse was compressed into a dense aluminum sandwich. Mr. Crow, running down the highway, was pitched end over end. Two cars belonging to the workers were wiped off the face of the earth. A pile of rubble that had been the clubhouse held nothing that could not be boxed up individually and sent away parcel post.

Ron Fountain, the airport base operator, was climbing into the rafters of his big hangar when the blast hit. He had a five-gallon pail full of parts in each hand and almost fell through the ceiling. At the time, he thought a student pilot had lost control and crashed his plane into the hangar—not an entirely rare event during that era.

Most of the large, west-facing storefront windows on Moscow's Main Street were cracked. Many residents initially thought the blast was an especially rigorous sonic boom caused by the many F-111s that trained over the scarcely populated region. Blaring fire sirens alerted them to the fact that they had just experienced something more intense.

Washington Water Power line crews working in the surrounding mountains remember the blast and the quasi-mushroom cloud it created. Residents of Troy, Idaho, eight air miles east—many of whom heard the blast—were sure it was nothing more than a sonic boom.

Mrs. Jorstad and her mother moved into a friend's house, where they stayed for more than three months. By the time they moved back, Mr. Crow was back in the pit with a new supply of ammonium nitrate, blasting out rock for his crusher like there was no tomorrow.

The Pullman Gun Club joined the Colton Gun Club rather than try to rebuild. Eventually the Pullman clubhouse was torched.

In spite of some intellectual veneer that might suggest otherwise, the incident was mostly forgotten by the community, basically at ease with explosives, firearms, and the outdoors. The local paper ran two news features and one mild editorial calling for tighter control of explosives, and then elected to drop the matter. Twenty years later, neither the now-retired chief of police nor most of the residents can recall the year in which the incident occurred.

Given the laid-back attitude of the residents and the sparse population, this ammonium-nitrate blast would probably not be classified as major. It was, however, large enough to catch people's attention for at least thirty days, till everyone forgot about it for all time.

Massive accidental blast cripples Soviet navy, kills 200

SEVEROMORSK, U.S.S.R., May 1984—Some very large explosions have occurred about which most of the world remains ignorant. Undoubtedly, huge numbers of witnesses exist, but due to the nature and location of the events, those witnesses will probably never come forward with details. The blast at Severomorsk, U.S.S.R., was that type of event.

On Friday, May 13, 1984, the Soviet military evidently endured the largest single conventional blast since World War II. A CIA analyst commenting at that time said it was "really a biggie." The explosion was so powerful that British and American intelligence experts initially thought they were looking at some kind of nuclear accident. It may have been even more powerful than anything previously encountered by the Soviet military, including wartime incidents.

Severomorsk lies on the north edge of the Kola

Peninsula at seventy degrees north latitude. It is not a city that appears on very many Western maps. Reliable sources estimate that about 55,000 people make their permanent homes there, the majority work in support of the military effort. The city actually lies north of Finland, closest to the thin section of Norway that wraps around Sweden and Finland, just under the Arctic Circle.

Severomorsk is the main base for the immensely powerful Soviet northern fleet. The fleet is charged with the delicate duty of protecting the polar seas and the North Atlantic for Mother Russia. Conditions are abnormally grim in this stark, barren part of the world. The Barents Sea—on which the Soviet fleet is forced to operate—is ice-clogged throughout all but a few months of the year. Raw rock knolls face out to sea in a monotonous pattern of blacks, grays, and deep browns. Little vegetation can survive in this atmosphere.

Murmansk is the nearest city with which non-Soviets might be familiar. Western equivalents include cities such as Barrow and Wainwright, north of Nome, Alaska. These are truly grim, difficult portions of the world—even without catastrophic explosions.

At the time of the blast, the Soviet northern fleet—composed of 148 surface ships of all sizes, from small aircraft carriers to light, Kotlin-class destroyers; 190 submarines of all classes; and 425 warplanes—had just completed the most extensive Soviet war games to date. Ammo and fuel stocks must have been at record highs. It was estimated that damage from the blast crippled the Soviet navy's ability to carry out their defense mission adequately for at least two years. Experts believed it would take at least that much time to rebuild the facilities and to manufacture and transport new supplies.

Although the blast apparently wiped out two thirds or more of the fleet's ordnance, there is no indication that any of the vessels themselves were damaged. Fortunately, a nuclear storage area just one mile from the depots was not involved. Of the six bunkers in the region, three were extensively damaged. Some of the storage areas contained nonexplosive vessel and plane spare parts, some of which may have been salvaged by the ant-like persistence of the many military personnel in the region.

A few days after the blast, Soviet commentators went to great lengths to explain that the explosion was not actually nuclear (a rather odd situation for the rest of the world, 99 percent of which resided at least one thousand miles from the blast site and could not have suspected that any kind of blast—nuclear or nonnuclear—had occurred).

Although the Soviets have never released official details to the West, we know from satellite photos and seismographic recordings taken from five hundred to one thousand miles away in Norway and Sweden that a huge blast occurred. Other "nonconventional" sources of information are alluded to but not confirmed in official reports. Eventually, Russian expatriates in Canada, Israel, or the United States may come forward with complete details. Perhaps insufficient time has elapsed, but as yet, no witness has stepped forward to provide details.

We know that a series of explosions continued on in a blasting, burning chain reaction for a minimum of five days. Greasy, thick smoke clouds from the blast were sufficiently intense to cause unusual atmospheric and climatic effects. Low-resolution weather satellites revealed that regional weather patterns were disrupted for several days afterward. The Western world realized shortly after the accident that something was amiss, but any sort of coordinated reports were painfully slow in coming. American news magazines did not report the incident until the week of July 2, fully six weeks after the fires had died, leaving ashes and millions of square meters of rubble.

Eventually the Soviets were forced to release a few details confirming that an incident had occurred, but the principal account of what happened came from an English defense magazine, *Jane's Defence Weekly*. *Jane's*, probably the most prestigious publication of its type in the world, discovered—perhaps through private, paid, on-site agents—that a minimum of two hundred people were killed and that 580 of the fleet's 900 SAN-1 and SAN-3 missiles, 320 of its 400 SSN-12s and SSN-3s, and the complete inventory of 80 SSN-22s were destroyed. Thousands of tons of bunker fuel, 20mm and 40mm cannon ammo, and 76mm and 127mm main-gun ammo reportedly were lost as well. *Jane's* also reported the loss of "some" SAN-6s and SAN-7s. (The SAN is the Soviet designation for SAM, or surface-to-air missile.) Given the use level, "some" in this case could have been thousands.

Soviet SSN-3 missiles were antiques at the time of the Severomorsk explosion. They were the first cruise missiles deployed by the Soviets. It had traditionally been Soviet policy to keep all equipment until it either completely deteriorated or they managed to sell it to a client state. That the SSN-3s were in storage in large numbers, therefore, is perfectly logical.

The loss of the modern SSN-12s, however, was definitely a blow to national security. The SSN-12, a five-thousand-pound submarine-launched missile with a range of about five hundred kilometers, was customarily used against enemy vessels. An SSN-12 carries an estimated four-hundred-pound warhead; these warheads may have been in place at the time of the accident.

SSN-22s were designed to carry nuclear warheads; the loss of all eighty of these—their most modern and expensive missiles—certainly left the

Soviet command feeling very exposed.

Records do not indicate exact numbers, but apparently many SAN-6 and SAN-7 antiaircraft missiles were also lost. It is possible that the Soviets stockpiled thousands of these missiles; therefore, the exact number on the day of destruction is unknown. With a kill ratio of approximately one aircraft downed for every fifty missiles expended, the captains probably had many hundreds, if not thousands, on hand.

Numerically, the greatest loss could have occurred among the SAN-1s and SAN-3s. The 1s were army antiaircraft missiles adapted for naval use. SAN-3s were more advanced models, but by 1984, even they were nearly obsolete. The claimed loss of 580 out of 900 total was probably not strategically serious except perhaps in the eyes of the pack-rat-like Soviet high command.

Back in those pre-glasnost times, estimates of total tonnage detonated over those fateful five days in May were by nature speculative. But because the incident spanned almost a week, it is not unreasonable to assume that tens of thousands of rocket engines, warheads, cannon, and main-gun rounds were involved.

Because the blast was so large that it was initially thought to be nuclear in nature, it is reasonable to speculate that the total amount of high explosives involved was at least three thousand to four thousand metric tons—other combustibles such as solid-fuel rocket engines and huge tanks full of Bunker C notwithstanding. The thousands of tons of Bunker C stored in the region could have conceivably taken much longer to burn, had the military not made stupendous efforts to minimize the damage.

Since the incident, blast-door security at the munitions bunkers reportedly is handled with great care. Standard safety practices that Americans would recognize immediately, such as one-way traffic regimes, no smoking, no drinking, and no sparking materials, are now commonly enforced. Apparently, enforcement of any safety standards was unusually lax on Friday, May 13, 1984.

Anonymous informants reporting to *Jane's Defence Weekly* said that fighter pilots were off-loading munitions from recently returned destroyers and hauling them by truck to "ready-use" magazines. Traffic was unusually heavy. Speculation has it that a truck loaded with missiles collided head-on with a returning missile carrier, setting off the first blast in the chain. (Even Soviet fighter pilots often do not know how to drive a vehicle. Driver training for peasants is often undertaken by the military services themselves. Noncombat casualties among drivers often reach forty men a day per hundred thousand deployed.) Once started, the blast apparently propagated from one bunker to the next through improperly closed blast doors.

Certainly, thousands of Russians, GIs, and workers who survived have a reasonably good concept of what caused the massive blast. Criminally tardy, superficial government reports attributed it to "too many munitions stored too close together." (A curious explanation at best, since—contrary to usual Western assumptions—the Soviet military is extremely cautious about handling explosives. Regard for human life—not always paramount in Soviet military thinking—is not the principal reason for exercising such care with explosives. The Soviet military assumes its mission of protecting the Motherland with extreme seriousness. Its strategists are chronically paranoid that they will not have sufficient war materiel with which to wage a credible action. Their caution might also be attributed to the fact that modern ordnance, regardless of the country in which it is manufactured, is breathtakingly expensive; like everything else in Russia, these supplies are exceptionally dear.)

Compared to Western experience with similar munitions and petrochemical explosions, two hundred dead seems to be something of an underestimate. Perhaps glasnost will eventually shed additional light on the incident, producing a more realistic casualty figure. Until then, based on seismographic readings and satellite photos, it will be regarded as the largest purely accidental military explosion since World War II, and perhaps of all time.

PEMEX gas plant explosions rip through Mexico City suburb

MEXICO CITY, MEXICO, November 1984—Travelers who have spent any time in Third World and other overpopulated countries know that the laws of nature frequently are repealed. In these places, people often behave as if the normal ground rules with which humans must generally comply or risk suffering instant, brutal consequences, simply do not apply. Often, they get away with it.

In Thailand, for instance, one can see a 440-volt

cable supplying power to an entire shopping center lying in the road exposed to everything from sixteen wheelers to motorcycles. Even when monsoon rains once backed water up three feet over the cable, no one was killed.

Korean construction workers customarily manufacture their own acetylene gas from carbide and water and store it in an old rubber inner tube. When they use their acetylene torches, they sit on the inner tube in an attempt to build up sufficient pressure to keep the torch going.

Ecuadoran delivery boys carry impossible loads of rolls and bread from bakery to consumer on their flimsy bicycles. One wonders as they zip through traffic what keeps the goods on the platter and how the riders ever manage to get to their destinations without spilling the precariously balanced load.

Somali Africans attempt perilous two-hundred-mile journeys through bunch grass desert with no thought of carrying water. Where normal rules prevail, they would certainly die horribly after twenty-five miles.

Usually, however, the laws of physics and nature eventually assert their authority, as they did in Mexico in November 1984.

Residents in the San Juan Eixhuatepac suburb of Mexico City reported that they had often detected the distinct odor of liquefied petroleum (LP) gas in their neighborhood near the Petroleos Mexicanos (PEMEX) compressor farm. The plant was one of thirty in the

MEXICO CITY, Nov. 20—SITE OF EXPLOSION—On page 155, fire fighters pour water onto two liquid-gas storage tanks as they burn following explosions at the PEMEX processing complex on the northern outskirts of Mexico City. Firemen in the foreground cool smaller storage tanks. Other exploded tanks lie akimbo in the background. (Photo courtesy of AP/Wide World Photos.)

sprawling Mexico City region. (The Mexican government decided during the 1930s and 1940s that rather than delivering the country's natural gas by pipeline, they would compress it and deliver it in twenty-four-gallon steel cylinders instead. As a result, this source of energy became more cheaply available to more people in more places than any other. Mexicans adapted wonderfully to the use of LP gas. Millions of liters were sold each month. Virtually every home had a gas stove. Few housewives—or even businessmen—could envision life without LP gas. Thousands of delivery trucks throughout Mexico left hundreds of storage and compression facilities such as the one in San Juan Eixhuatepac with their vital loads of fuel each working day.)

Ideally, homes should not have been constructed on the land immediately surrounding the PEMEX gas plant. But this was Mexico City, where undeveloped land was scarce and paid-off officials were extremely common. Thus, there were an estimated 150,000 semi-impoverished people living in the immediate area in adobe and tin buildings. Most of the homes had their own LP gas cylinders outside their kitchen windows. The material was common, and therefore perhaps not given due respect.

As an explosive, LP gas is a bit strange. It will not detonate at concentrations of less than 4.5 percent or greater than 12 percent gas-to-oxygen-ratio. Perhaps the gas smell residents detected in the neighborhood for the preceding several months was below the maximum concentration, or even above the minimum. (The latter seems unlikely, but one must remember that normal rules of physics don't always apply.) Some residents reported the unique gas odor to authorities, but apparently no one took the reports seriously enough to do anything about them.

At 5:42 A.M. on Monday, November 19, 1984, gas

concentrations apparently reached the critical stage and exploded with a resounding whomp that could be heard at least twenty miles across the huge city. Even properly mixed LP gas is not a sharp, fast explosive. It takes large quantities to detonate with enough force to be heard over any measurable distance.

The initial blast was a series of four explosions as four 420,000-gallon tanks of LP gas went up. Because Mexico City was originally built on a series of lakes, even far-off residents of the city could feel the impact through the mashed potato-like ground. Steel shards from the massive steel tanks were thrown about the area. Rather than creating the mushroom cloud that commonly accompanies huge blasts, this well-fueled explosion created its own fire storm.

Survivors estimated that flames leapt three hundred feet or more into the sky. The huge fire storm that settled on the locale like a great fog wiped out thirty acres of squatters' buildings. Most of the houses were constructed of cement blocks and had cement floors and tin roofs. If the firestorm had settled on a poorer section built mostly of cardboard and scrap, the situation might have been even more desperate. As it happened, buildings were shattered, much like in a conventional blast. All of the roofs in the immediate impact area were blown away, as were those of many of the homes in surrounding regions. Heavy masonry walls were warped, cracked, and toppled. Rubble lay everywhere in the streets and alleys.

In spite of fairly extensive probing by Mexican

MEXICO CITY, Nov. 20—SCENE OF BLAST—Two liquid-gas storage tanks after explosions ripped through a Mexico City suburb, leaving hundreds dead and injuring scores of others. (Photo courtesy of AP/Wide World Photos.)

authorities, the exact cause of the blast was not determined. Testifiers at an investigative hearing claimed that a truck at one of the private compressing plants crunched into a loading dock, producing a shower of sparks that initiated the blast.

Owners of that private compressing concern were quick to point out that the only tanks that exploded were those owned by PEMEX and located on PEMEX property. Since the driver of the truck was reported as a probable casualty in the initial explosion, the authorities allowed that entire avenue of questioning to drop into the proverbial black hole.

Alberto Aquino Hernandez was on the PEMEX site counting twenty-four-gallon cylinders for the delivery truck when the blast occurred. At the first sound of the explosion, he ran downhill away from the enveloping fire storm until he was out of the burn area. Mr. Aquino thought the explosion originated on the PEMEX site, but he did not know for certain by what means.

He was one of the very few in the impact area to escape. One fifty-foot cylinder weighing a quarter ton blew itself half a mile away toward the heart of the city. In addition to the four main steel tanks containing more than 1.5 million gallons of LP gas, forty-eight smaller tanks also detonated. The larger tanks were each the size of several box cars and made of very heavy plate steel.

An estimated 100,000 people evacuated the area, managing to escape the sheet of rolling flame that passed through, instantly carbonizing an estimated 540 unlucky souls. Another 2,000 were seriously injured. As with most explosions, the casualties occurred very quickly, perhaps inflicting little pain. The real distress occurred among the survivors, who looked futilely for relatives and friends among the reported 1,000 missing.

It was perhaps appropriate that authorities elect-

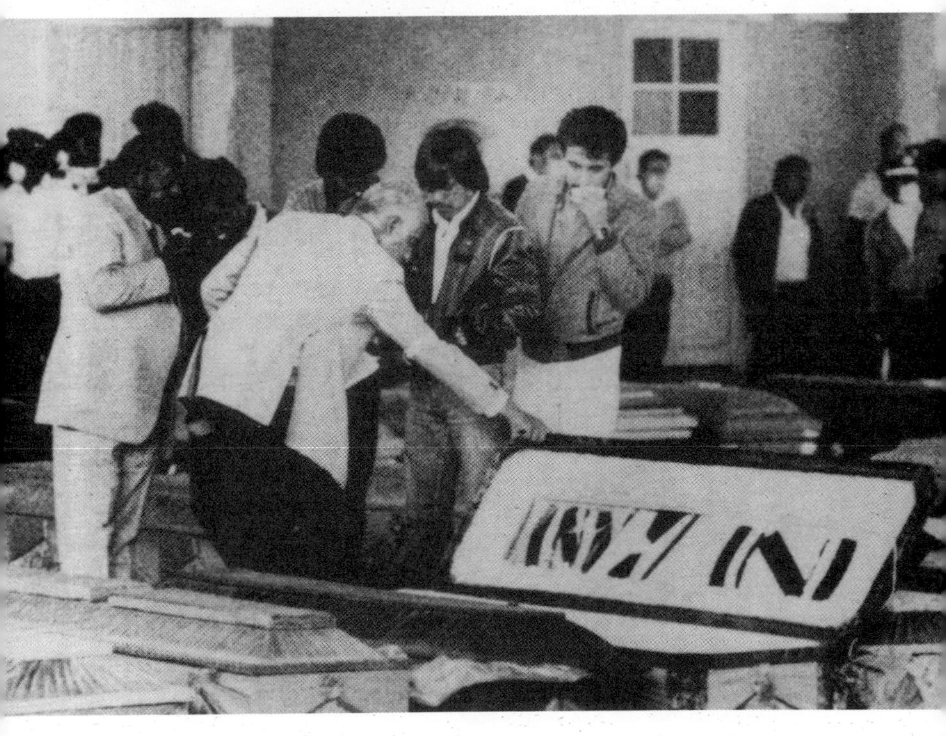

SANTA MARIA TULPETLAC, MEXICO, Nov. 20—IDENTIFYING THE DEAD—At a makeshift morgue in this Mexico City suburb, survivors search for loved ones killed in what ranks as one of the country's worst disasters. Initial reports from the Red Cross stated that nearly 300 bodies were recovered and that at least 500 people were seriously injured. In the end, the blast left 540 dead, 1,000 missing, and 2,000 seriously injured. (Photo courtesy of AP/Wide World Photos.)

ed to level the surrounding area, clearing it of all habitation. As a safety measure, reducing the number of squatters' buildings was laudatory. Even the more permanent cement block and tin buildings were not rebuilt after the bulldozers left. Any piece of open land in a place such as Mexico City, however, is extremely attractive to people with very little, no matter how dangerous it might be.

The new residents no doubt hope that PEMEX will keep its fittings tight and not tempt the laws of nature and fate again by setting up conditions that could lead to a repeat performance of what was probably the worst gas explosion in history.

EDITOR'S NOTE: Gas explosions unofficially killed more than one hundred people in an apartment complex in Tbilisi, U.S.S.R., on December 2, 1984. No official reports document this explosion; thus, the Mexican gas blast is still considered the greatest of all time.

Foul play suspected in bombing of General Li's mansion

CHIANG MAI, THAILAND, March 1984—Harlan Lee, the highest ranking American diplomat in northern Thailand, was away on business at the time of the blast. His wife, however, was in their giant wood-frame home within the brick-walled consulate compound attending to the needs of their new baby.

Partly because of the wall circling the compound, the blast did little more physical damage than to crack all the north windows. Mrs. Lee remembers that it was powerful enough to give the house a "good shake," however. She recalled struggling to hold the baby as she was nudged to the floor. The incident was threatening only in terms of its suddenness. The impact, she recalls, was terribly surprising.

Mrs. Lee listened intently for the telltale roar of jets, thinking that the blast was little more than a sonic boom. For a few minutes there was nothing but the hush of the city preparing for nightfall. Then, in the distance, sirens started to howl. Soon she heard the fire trucks departing from the main station across the street from the consulate. She listened as the trucks continued up the road along the Ping River toward the origin of the blast.

The trucks headed into a wealthy section of Chiang Mai, originally settled and built up by missionaries some fifty years earlier. Now it was a mostly flood-free enclave, inhabited by well-to-do Thais and people receiving a regular check from churches in North America and the United Kingdom.

Very quickly, other excited sounds joined the cacophony building in an otherwise laid-back area of the Orient. Mrs. Lee knew something was desperately amiss in this, the most fabled city in the world. After waiting discreetly, since she assumed the matter was probably not of American concern, she placed her baby in the hands of her number-one ser-

CHIANG MAI, THAILAND, Mar. 12—LI'S MANSION IN SHAMBLES—Shown on page 163 is the residence of the notorious and elusive General Li following a mysterious explosion that occurred in the general's absence. The blast, which originated in a van parked inside the closely guarded compound, leveled a garage, wrecked two Mercedes and a Toyota truck, tore off the front of the house, and destroyed the kitchen. Surrounding structures in the neighborhood also suffered extensive damage. While reports of casualties varied, official counts had one person dead and twelve injured.

vant and walked to the compound gate.

The consulate guards and Thai police gathered at the gate told her an explosion had gone off in the courtyard of a rich man's home east of the river. Little more was known about the incident at the time. Only slightly more solid information has filtered out since.

Although it was not commonly known to residents even in the immediate neighborhood, the palatial home and grounds hit that evening constituted the residence of the notorious and elusive General Li. At the time, Li was reportedly in his early eighties. Originally an immigrant from China, he came to Thailand in 1949 after the fall of Chiang Kai-shek, settling on the Burmese border with his ragtag little army.

Although the Thais were not overjoyed at his presence, they allowed the Chinese army to remain in the area as a counterforce to private drug armies and communists operating in the region, who continued to pose a threat until the early '70s. By the mid '80s, the Thais could have thrown General Li out, but elected to ignore the situation because of his money and power.

General Li's Kuomintang (or KMT) party maintained a fairly high profile in that region of the Golden Triangle. But Li himself dropped out of sight. He became so elusive and obscure that few Thais knew or were even familiar with the man by sight. The only means the average Thai citizen had of knowing of General Li's existence was through gossip or occasional contact with one of his few legitimate businesses. Li himself, it was reliably rumored, made the principal portion of the considerable income that supported both himself and his army by trafficking in illegal narcotics. The principal crop in the northern mountains of Thailand was opium poppies, a situation on which he capitalized to a great extent.

Through the years, General Li, supported by a better-than-average business acumen, organized many of the Burmese and Thai ridgetop heroin refineries. Eventually, he controlled production facilities for considerable amounts of highly condensed No. 4 heroin, which he wholesaled to mobsters in New York, Hong Kong, Manila, Amsterdam, and many other places where concentrations of addicts existed.

General Li's ruthless movement into the heroin trade earned him a number of equally ruthless enemies. As is true the world over in the narcotics trade, his activities were characterized by frequent ongoing turf wars. If it was true, as initially supposed, that his competition was out to get him, the enemy's intelligence was only accurate insofar as the exact location of the house was concerned. Everything else was done so poorly that many authorities speculated that it was an accidental detonation.

At the time of the blast, Li was not even at home. The bomb accomplished little except to trash the immediate neighborhood. To do maximum damage to Li's house, it should have been pointed in the opposite direction. Whether General Li ever returned to his bent and crippled home in Chiang Mai remains unknown.

Theories about what really happened that Monday evening abound. The truth will almost certainly never come out. In all probability, every one of those concerned with the blast is now dead or has left the country. Only the big bosses remain, and they, of course, do not answer calls from journalists asking for interviews.

The explosion originated in a van parked in front of a garage inside the closely guarded compound. How a van carrying a heavy load of explosives managed to pass through tight gate security into the inner court of the grounds has never been explained. It appears as though the van may have belonged to the general. Since no one has stepped forward with precise information, estimates of the amount of explosives involved are pure speculation.

Upon detonation, the blast completely leveled the garage, badly crippled two relatively new $300,000 Mercedes and a new Toyota truck, tore off the front of the columnar, antebellum-style house, and destroyed a large, free-standing kitchen. Tall palms sheltering the gardens were stripped and shredded. Two days later, the shrubbery still hung in tattered shreds, flapping grotesquely in the wind.

Casualties at the American school located just to the northeast across the alley could have been extensive had the blast occurred several hours earlier, but none of the more than five hundred students were in the building at the time. However, the entire auditorium facing Li's house was blown in.

In addition to wiping out the smaller outbuildings on General Li's compound, the blast reduced a small house immediately to the east and one across the street to disorganized piles of rubble. To some extent, the heavy wall around the compound shielded the lower stories of houses on surrounding blocks, yet entire second and third stories of many were whooshed away by the impact.

Down the block, a wealthy Thai lost several hundred square meters of red-clay roof tiles. Without homeowners insurance (virtually unknown in that part of the world), and with massive spring rains on the way, property owners had little recourse but to dig into their own pockets expeditiously and make their own repairs.

A huge, ornate Buddhist temple a block away suffered extensive damage to its roof and golden gingerbread decorations. For the devout, gentle Thais, this violent intrusion constituted the ultimate act of sacrilege. Scores of volunteers undertook repairs immediately and reverently.

In spite of the obvious destruction to the tightly

packed neighborhood, most residents had little idea what had really transpired. Bordering property owners maintained that they knew little about their reclusive neighbor. For the vast majority of Chiang Mai residents who flocked to the scene, it was the first proof they had that General Li really lived among them.

Police estimates indicated that the blast was caused by three hundred to five hundred pounds of common, low-yield dynamite. While strapping the dynamite to LP gas cylinders may have intensified the blast somewhat, it was not particularly massive. If the same detonation had occurred in a less densely packed area, the effects might have been inconsequential. In a country where repair costs and skilled labor were still reasonable in the 1980s, property damage was estimated at roughly one million U.S. dollars.

General Li was also heavily involved in the Burmese jade trade, which involved smuggling commercial explosives in for the miners, who bartered with the raw jade and opium that Li smuggled back to Thailand. Therefore, initial police speculation suggested that one of his own powder magazines accidentally detonated.

The theory that the blast may have been accidental gained credence when investigators discovered no evidence of renewed warfare between regional drug lords. This, as well as the strange presence of the van and General Li's absence from his home at the time of the blast, strongly suggests that the explosion was unintentional.

Official police records show one casualty and twelve hurt. Police who were working in the area of the blast, however, claim—unofficially—that at least one bathtub salvaged from the wreckage was filled with human body parts. Unofficial sources claimed a driver, cook, gardener, and several others were never seen again after the blast.

Although no official report to this effect was ever released, American authorities have come to believe that the blast was, in fact, an amateurish attempt to take General Li out with a car bomb using commercial dynamite. Intelligence gathered by U.S. narcotics agents seems to suggest that the blast was ordered by Khun Sa, the notorious Shan United Army chieftain, as a means of humiliating or killing his chief rival.

Even though neither General Li nor any of this key people were physically injured, he was humiliated. For the first time, thousands of prying eyes looked in on his private compound. For the first time, everyone knew he actually lived among them. His reputation suffered extensively in the eyes of those he led when it became obvious he could not protect his own home.

His loss of face certainly had a much greater impact than the explosion itself. But still, accidental or intentional, it was a devastating event for that particular time and place.

Index

N

O

P

R

S

Y

Z